UNDERSTANDING TIME AND SPACE

An invitation to the theory of relativity
for anyone who is now, or has ever been,
an inquisitive high school student.

UNDERSTANDING TIME AND SPACE

An invitation to the theory of relativity
for anyone who is now, or has ever been,
an inquisitive high school student.

Steven E. Landsburg
University of Rochester, USA

World Scientific

NEW JERSEY • LONDON • SINGAPORE • BEIJING • SHANGHAI • TAIPEI • CHENNAI

Published by

World Scientific Publishing Co. Pte. Ltd.

5 Toh Tuck Link, Singapore 596224

USA office: 27 Warren Street, Suite 401-402, Hackensack, NJ 07601

UK office: 57 Shelton Street, Covent Garden, London WC2H 9HE

Library of Congress Control Number: 2025931740

British Library Cataloguing-in-Publication Data
A catalogue record for this book is available from the British Library.

UNDERSTANDING TIME AND SPACE

ISBN 978-981-98-0839-7 (hardcover)
ISBN 978-981-98-1310-0 (paperback)
ISBN 978-981-98-0840-3 (ebook for institutions)
ISBN 978-981-98-0841-0 (ebook for individuals)

For any available supplementary material, please visit
https://www.worldscientific.com/worldscibooks/10.1142/14189#t=suppl

Typeset by Stallion Press
Email: enquiries@stallionpress.com

Preface

How and Why This Book Will Help You Learn About Relativity

My hobby is answering questions about relativity on the Internet. The questions come from high schoolers, college students, lifelong learners, and now and then a professional physicist who has managed to get confused about something basic. (It happens to all of us.) I get a lot of feedback, including thank-you notes, followup questions, and occasionally clear evidence that I've been thoroughly misunderstood.

The good news for you, as a reader of this book, is that all that feedback has taught me a lot about what works, what doesn't, and how I can be most helpful. Here are some examples:

- **Relativity can seem violently counterintuitive:** Fundamentally, it says that different observers can describe the world in very different ways and they can all be correct. This is less remarkable than it sounds. After all, you've known since childhood that sometimes the direction you call left might be the same as the direction I call right — and we can both be correct.

  This book will illuminate and demystify the theory of relativity with repeated analogies to the everyday and the familiar. The first five (very short) chapters are devoted entirely to a brief and straightforward parable about people giving each other directions on a city street. Then, as we move on to the substance of relativity theory, I'll frequently remind you of that simple parable to ground your sense of what's really going on.

- **Learning relativity requires clear pictures:** That's because relativity is a branch of geometry — the geometry of seeing the world differently from different perspectives. Unless you have a very unusual sort of brain, you can't learn geometry without pictures.

  This book contains a lot of pictures. At first glance, some of them might look pretty complicated, and maybe even scary. Don't panic. I'll carefully talk you through the meaning of each individual dot and line in every one of those pictures, and once you've worked through the explanations, you'll be surprised how simple they start to look.

  Most of the pictures in this book have been field-tested on the internet, where I've posted them to illustrate my answers. Sometimes, I get great feedback like "Oh! That picture makes everything clear now!". Other times, I get less gratifying feedback like "I still don't quite get it", or "What is that red line doing there?" or even just "Huh?". Then I redraw the picture, collect more feedback, and if necessary redraw it again. The versions that made it into this book are the ones that have been redrawn as many times as necessary to make people say "Oh! *Now* it's all clear!"

  All books about relativity contain some pictures. But many books treat the pictures as second-class citizens, emphasizing words or equations instead. I've found that students who have learned from those books often share a lot of the same misconceptions.

- **When students go astray, it's usually because they've latched on to one or another of a small number of common misconceptions:** That's great news because it means that I can ease your path by warning you about those misconceptions before they get a chance to trip you up.

  Throughout this book, you'll find short sections set off with the dangerous curve symbol:

This symbol is there to warn you that this is the exact place where many students go off the rails by misunderstanding something important. In the main text, I'll explain why the right ideas are right; in the dangerous curve sections, I'll explain why the wrong ideas are wrong.

- **It pays to work problems:** There's no better way to fix an idea in your head than to apply it. There are exercises sprinkled throughout this book and detailed answers to all of them at the end. You won't find them at the ends of chapters; instead you'll find each exercise at the exact location where you first ought to be able to solve it (or at least to try). Whenever you encounter an exercise, I recommend pausing to tackle it before you move on.

- **Not everyone wants to see the same details:** For example, a little trigonometry will probably clear the air for some readers while it stops others dead in their tracks. This book aims to be hospitable to everyone. So, the sections that require a little more background are marked with the star symbol: ✷. The book is written so that you can easily skip these sections and still understand everything that follows.

- **It pays to keep reviewing:** An idea that makes perfect sense in the morning can seem pretty hazy by the time you sit down to dinner. From time to time, you'll probably want to leaf backward and reread the key passage that temporarily made everything so clear. Unfortunately, the exact location of that passage can also be pretty hazy — was it five pages back or thirty?

  The detailed Table of Contents, including short summaries of each chapter, is designed to help. By scanning the contents, you can quickly locate the right chapter, and because the chapters themselves are all very short, I hope it will then be quite easy to find exactly what you're looking for.

Finally, if you learn relativity from this book, you'll owe a debt of gratitude to everyone who has given me feedback on earlier drafts. You can pay that debt forward by emailing me your own feedback at questions@landsburg.com. Thank you!

A Note to Readers Who
Have Already Studied Relativity

This book was written for beginners. But it can be enlightening for non-beginners as well. Sometimes, it helps to see the same ideas presented in more than one way. This is especially true of relativity, where a lot of the ideas take some getting used to.

For example, as a non-beginner, you're probably aware that sometimes Alice and Bob disagree about whose clock is running faster and relativity says they're both correct. Students (by which I mean anyone who is trying to learn about relativity, whether or not they're actually enrolled in school) often think they must have misunderstood this. How can such a thing possibly *be*?

Many teachers, many textbooks, and many physics sites on the Internet answer that question with appeals to thought experiments or mathematical derivations that *prove* (given a few basic postulates) that this is how things *must* be. Those answers are all perfectly correct, but can still feel unsatisfying. Worse yet, those answers represent a missed opportunity to convey a clear picture of what relativity is really all about.

This book takes a different approach. It will repeatedly ask you to put relativity aside for a moment and focus on something far more familiar. Charlie, standing in Chicago and facing north, says that Los Angeles is "to the left", while Dora, facing Charlie, says that Los Angeles is "to the right". You and I say they're both entirely correct. How can such a thing possibly be?

The answer (and I'm sure you're ahead of me on this) is that Charlie and Dora are facing *different directions* in space, and when you turn around in space, the directions you call "left" and "right" turn with you. *Understanding relativity means understanding that the answer to the Alice/Bob question is exactly the same as the answer to the Charlie/Dora question.* When Bob takes off in his rocket ship, he is *turning around*, not in space but in spacetime. Just as turning around in space forces you to adjust your notions of left and right, turning around in spacetime (i.e., changing your speed or your direction) forces you to adjust your notions of how fast various clocks are ticking. If you don't yet fully grasp that analogy, this book has something to teach you.

Another example: Accelerating your rocket ship means turning around in spacetime, which changes the way you describe the rest of the universe. When you accelerate, you might switch from saying *Earth clocks are showing a time of 1 PM right now* to saying *Earth clocks are showing a time of 10 AM right now*. That's because your acceleration changes your notion of "right now on Earth" in exactly the same way that your decision to face south instead of north changes your notion of which way is left.

Students who have learned relativity from traditional textbooks often miss this point completely. Instead, they manage to convince themselves (implausibly enough) that your decision to accelerate your rocket ship somehow causes all those clocks back on earth to jump backward by three hours — or, less implausibly but equally incorrectly, that it causes your own clocks to jump forward. The students who avoid this trap are those who have learned to think of relativity as the *geometry of changing perspectives*. They are, in other words, the students who have learned to think in pictures and to relate those pictures to their everyday experience. This book aims to make you one of those students.

Contents

Alice and Bob, standing on Seventh Avenue and facing each
other, attempt to point their friend Jeter to Carnegie Hall.
He is surprised that Alice says it's to the left, while Bob
says it's to the right.

We address Jeter's confusion by observing that Bob and
Alice have different *frames of reference*, and that there are
simple *coordinate transformations* that translate between
Alice's language and Bob's.

Although Alice and Bob describe some things in different
languages, there is at least one thing they describe exactly
the same way: The *distance* between any two locations.
We therefore say that the distance is an *invariant* of the
coordinate transformations.

Alice and Bob are now standing in the desert, facing different directions. Now they "disagree" about the meanings of
Forward and *Backward* as well as *Left* and *Right*. (Of course
they don't really disagree; they just use the same words to
mean different things.) Once again, we can write down the
coordinate transformations that translate between Alice's
language and Bob's.

If Alice and Bob are standing at different *locations* in the
desert (while continuing to face in different directions),
we can still use coordinate transformations to go back
and forth between their descriptions. However, we have to
apply these transformations to the *differences* between two
objects' x and y coordinates, rather than to the coordinates
themselves. As on Seventh Avenue, the *distances* between
various objects are invariants of these transformations.

Alice and Bob are back on Seventh Avenue, where Alice is
standing at the corner of 52nd sreet while Bob skateboards
to the east at 1/2 block per hour. Jeter asks to be pointed
to tonight's concert, which means he's asking to be directed
not just to a specific location but to *a specific location
at a specific time*. We introduce a graphical depiction of
spacetime, in which each event (such as the concert) is
represented by a single point. On this graph, Alice, Bob,
and Carnegie Hall are represented by lines, called their
worldlines.

Any observer whose worldline is straight is called an
inertial observer. The *First Principle of Relativity* says
that any inertial observer has the right to consider himself
stationary, and no experiment can prove him wrong. Guided
by this principle, we can construct a candidate for Bob's
frame — all of the "fixed location" lines are parallel to his
worldline, and all of the "fixed time" lines are, like Alice's,
horizontal.

Unfortunately, the frame we've constructed in the preceding
chapter runs afoul of the *Second Principle of Relativity*:
All inertial observers must agree that the speed of light
is exactly one light-hour per hour. But according to the
"Theory of Relativity" we constructed in Chapter 7, if Bob
is traveling eastward past Alice at speed 1/2 when she
shines a light beam eastward at speed 1, he will measure
the speed of the lightbeam as 1/2, not 1 as required by the
Second Principle. Thus we need a revised theory, which is
to say we need to reconsider Bob's frame.

Here's what works: If Bob is traveling at velocity v with
respect to Alice, then when Alice says an event occurs at
location x and time t, Bob says it occurs at location x' and
time t', where

$$x' = \frac{x - vt}{\sqrt{1 - v^2}} \qquad t' = \frac{t - vx}{\sqrt{1 - v^2}}$$

These *Lorentz transformations* allow us to translate back
and forth between Alice's and Bob's description of the
world.

Using the Lorentz transformations, we can depict the time
and space axes from both Alice's frame and Bob's on the
same graph. This allows us to discover that events can be
simultaneous in one frame but not in the other.

We add some grid lines (corresponding to fixed times and
fixed locations) to both frames and confirm that the speed
of light is the same in both frames, just as the Second
Principle of Relativity requires.

We stress the analogy between the Lorentz transformations
that translate between Alice's and Bob's descriptions of
spacetime, and the coordinate transformations that translate between Alice's and Bob's descriptions of the desert.
We also observe that — just as in the desert — when
Alice and Bob don't share a common origin, the Lorentz
transformations can be applied to *differences* between
coordinates, rather than the coordinates themselves.

When Bob travels from Alice to a distant location, they will
disagree about how much time passes along the way. We
use the Lorentz transformations to show that, according
to Alice, Bob's clock runs slow by a factor of $\sqrt{1 - v^2}$,
while according to Bob, Alice's clock runs slow by the same
factor.

Chapter 14   A Journey from the Stars

When Bob travels toward Alice from a distant location, we confirm that the conclusions of the preceding chapter still apply.

Chapter 15   A Round-Trip Journey: How Travel Keeps You Young

When Bob makes a round-trip journey from Alice to a distant location and back, we use the Lorentz transformations to calculate that upon his return, Bob has aged less than Alice has.

Chapter 16   A Birthday Party

We revisit Bob's roundtrip journey from the previous chapter, assuming that Alice livestreams video of herself, and calculate what Bob sees on his video screen over the course of the trip.

PART V. THE ORDER OF EVENTS

Chapter 17   What We Can All Agree On: The Spacetime Interval

We define the *spacetime interval* between two events, and confirm that the interval is an invariant of the Lorentz transformations — that is, even when Bob is in motion with respect to Alice, they will always agree on the interval between any two events. When an inertial observer travels from one event to another, the amount of time that passes on the traveler's clock is equal to the interval between the events.

When Bob is in motion with respect to Alice — say at velocity v — they will disagree about the distance from Alice to some other object (say a distant star) which is stationary with respect to Alice. In fact, we use the Lorentz Transformations to show that according to according to Alice, Bob underestimates the distance by a factor of $\sqrt{1 - v^2}$.

When Bob flies past Alice — once again at velocity v — they will disagree about the distance from Bob to some other object (say the front of Bob's spaceship) which is stationary with respect to Bob. In fact, Alice says that Bob overestimates the distance by a factor of $1/\sqrt{1 - v^2}$, and therefore overestimates the length of his ship by that same factor.

If Bob is at rest with respect to Alice, they must agree on the length of Bob's spaceship. After he takes off, we know from the preceding chapter that he must say his his ship is longer than Alice says it is. This raises the question: Does Alice say the ship has shrunk, or does Bob say the ship has stretched? In these two chapters, we'll see that either answer might be correct, depending on the details of how the ship takes off. This is a cautionary lesson for many other problems in relativity: Details matter!

A step-by-step guide to discovering the Lorentz transformation for yourself.

PART I
FRAMES IN SPACE

Chapter 1

How Do You Get to Carnegie Hall?

Alice and Bob are standing on a New York street corner facing each other, when they are approached by Jeter, who asks for directions to Carnegie Hall.

"It's five blocks to the left", says Alice.

"No, no! It's five blocks to the right!", says Bob.

Who should Jeter believe?

He should, of course, believe them both. Carnegie Hall happens to be both five blocks to Alice's left *and* five blocks to Bob's right.

To you, that's obvious. But Jeter has an odd mental block about such things. To Jeter, it seems clear that if Carnegie Hall is to the left, it can't also be to the right. He wants to know the *true* direction to Carnegie Hall.

"It's in the direction we're pointing!", says Alice. "Can't you see that we're both pointing the same way?"

"Yes", says Jeter. "But what direction *is* that? Is it left or is it right?"

At this point, Bob turns 180 degrees, so that he's facing the same direction as Alice. "Carnegie Hall is five blocks to the right", he says.

"Whoa!", says Jeter. "Just a moment ago, you said it was five blocks to the left! So, you've changed your mind?"

"No", says Bob. "The only thing I changed was my body. When my body faced one way, Carnegie Hall was five blocks to my left. Now, that my body faces a different way, Carnegie Hall is five blocks to my right."

"So let me get this straight", says Jeter. "You're saying that *just by turning yourself around*, you can make a gigantic concert hall shift its location by *ten blocks*? That's *crazy!*"

If you are encountering relativity for the first time, you are Jeter. You — like everyone who has not been trained in relativity — have some mental blocks. When different observers describe the same thing in different words, you're going to want to know who is correct and who is mistaken. When an observer, like Bob, changes his perspective and so changes his description, you're going to to suspect him of claiming that the rest of the Universe has shifted around.

Stay calm. If Jeter pauses to think this through, it will all make sense to him. And to you.

Chapter 2

Frames of Reference

Here is a map of Seventh Avenue in New York City:

The map shows Alice, pointing toward Carnegie Hall, which is to your right and her left. Bob, like you, is facing toward Alice, so Carnegie Hall is to his right also.

Here is a simplified version of that map, with Alice replaced by a black dot, and with circular markers at every block:

If Bob labels the markers according to how far rightward they are, he'll draw this picture:

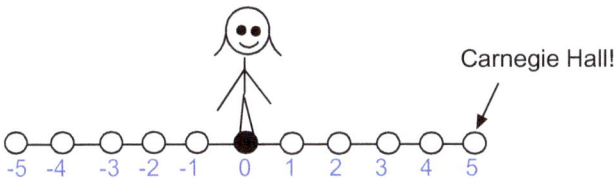

This assignment of numbers to locations is called Bob's *coordinate system* or his *reference frame* (these phrases mean exactly the same thing). For example, Bob's reference frame assigns the number 0 to Alice and the number 5 to Carnegie Hall. Sometimes, we abbreviate the phrase *reference frame* to the single word *frame*.

If Alice labels the markers according to how far rightward they are, she'll draw this picture, representing her own reference frame:

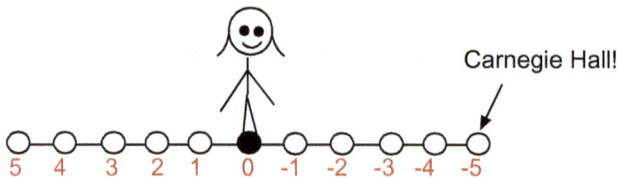

Here's the picture again, with both Bob's and Alice's reference frames shown:

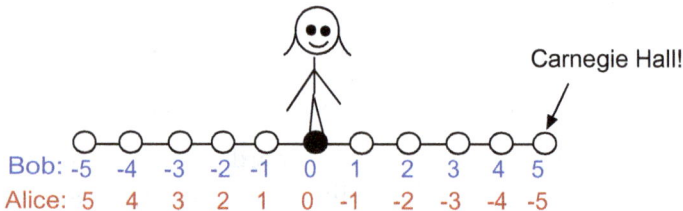

The numerical labels in this picture are called Alice's and Bob's *coordinates*.

The point that Alice calls 3 is the same as the point that Bob calls -3. The point that Alice calls -2 is the same as the point that Bob calls 2. In fact, there's a quite simple formula relating Bob's entire reference frame to Alice's entire reference frame. Namely, wherever Alice assigns the coordinate x, Bob assigns the coordinate $-x$.

Or to put this a slightly different way: Pick any circular marker. Let x represent the coordinate Alice assigns to that marker, and let x' represent the coordinate Bob assigns to that marker. Then we have these formulas

$$x' = -x \qquad (2.1)$$

$$x = -x' \qquad (2.2)$$

These formulas, which allow us to go back and forth between Alice's frame and Bob's, are called the *coordinate transformations* between those frames.

Note that the coordinate transformations reflect a fundamentally *geometric* picture. To get from Alice's frame to Bob's (or vice versa) you can imagine taking that frame and *reflecting it* through the point where Alice is standing. The equations describe the result of that reflection.

Chapter 3

Invariants: What We All Agree On

Now let's look at a slightly more detailed map of Seventh Avenue, showing a few more landmarks:

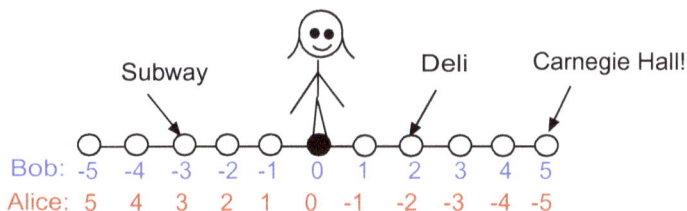

Jeter has just asked Bob how far it is from the Deli to Carnegie Hall. Bob can use his (blue) reference frame to calculate the answer: The Deli has coordinate 2, Carnegie Hall has coordinate 5, and the distance between them is the *difference*, namely $5-2$. It's three blocks from the Deli to Carnegie Hall.

How did Bob know to subtract 2 from 5 and not the other way around? Because distances are always positive. He could perfectly well have subtracted 5 from 2, but then he'd have to report the *absolute value* of his calculation.

More generally, if Bob wants to know the distance between two points with coordinates x_1 and x_2, he can compute that distance via the formula

$$|x_1 - x_2|$$

Using this formula, the distance from the Subway to the Deli is $|(-3) - 2|$, otherwise known as 5. The distance from Carnegie Hall to the Subway is $|5 - (-3)|$, or 8.

Alice can do exactly the same thing using her (red) coordinates. The distance from the Deli to Carnegie Hall is $|-2-(-5)| = 3$. The distance from the Subway to the Deli is $|3-(-2)| = 5$. The distance from Carnegie Hall to the Deli is $|(-5)-(-2)| = 3$.

Although Alice and Bob are using different coordinates, they keep getting the same answers. We express this by saying that the distance between two landmarks is an *invariant* of the coordinate transformations. I hope it's obvious to you that the distance really is an invariant — in other words, that *for any landmarks at all*, the distance computed by Alice will always be the same as the distance computed by Bob. This is exactly what it means for distance to be an *invariant*.

✱

> Boxes marked with the ✱ symbol provide mathematical details that you can safely ignore.

Although it might be obvious that distance is an invariant, it can be worth checking to make sure. So:

Suppose we have two landmarks, to which Alice associates the coordinates x_1 and x_2, and Bob associates the coordinates x'_1 and x'_2. Then Alice computes the distance between those landmarks as

$$d_A = |x_1 - x_2|$$

while Bob computes the distance as

$$d_B = |x'_1 - x'_2|$$

(continued on next page)

(continued from previous page)

But the coordinate transformation (2.1) tells us that $x_1' = -x_1$ and $x_2' = -x_2$. So we can rewrite Bob's computation as

$$d_B = |(-x_1) - (-x_2)|$$
$$= |-(x_1 - x_2)|$$
$$= |x_1 - x_2|$$
$$= d_A$$

In short, $d_B = d_A$. Bob's and Alice's computations are different, but they agree in the end. To repeat, this is exactly what it means for distance to be an *invariant*.

Chapter 4

Lost (and found) in the Desert

Alice has left New York and is now standing in a vast desert. This desert differs from Seventh Avenue in many ways, but we'll focus on two of them. First, unlike Seventh Avenue, which is essentially a line, the desert is essentially a plane. In other words, we've added a dimension. Second, unlike Seventh Avenue, which has cross streets at one-block intervals, the desert has no such natural markings. It looks like this:

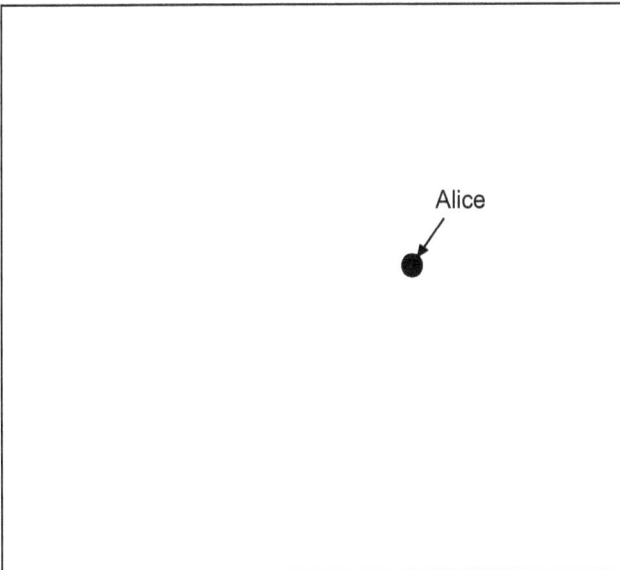

Not only are there no natural markings in this desert; there are no natural directions either — that is, there is no way for Alice to distinguish north, south, east or west. (If you find that hard to believe, just imagine that Alice is particularly direction-challenged, or, if you prefer, that the desert is floating in space.)

Nevertheless, Alice has no problem constructing a reference frame. She certainly *does* know the difference between forward, backward, left, and right, and she can measure distance by the yardstick she always carries. So, her reference frame looks like this (Alice happens to be facing the top of the map):

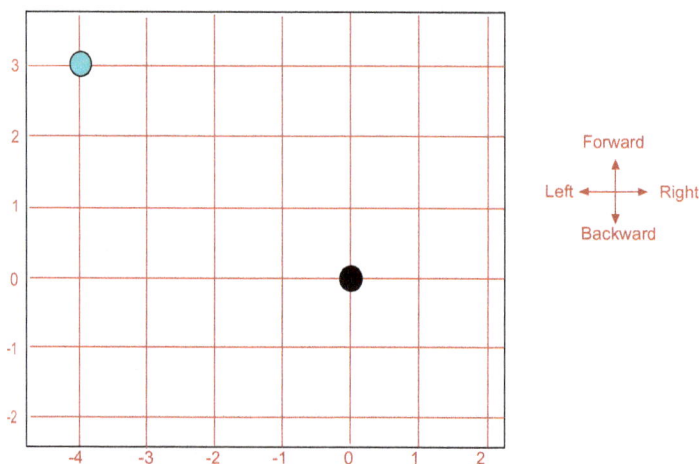

The red lines and labels, of course, exist only in Alice's head. But they can help her keep track of important realities, like the location of the green tree where she left her lunchbox — it's 4 yards to the left and 3 yards forward, at the point with coordinates $(-4, 3)$.

Now, Bob has arrived and is standing right next to Alice — but he's facing a different direction — in fact in the direction that, according to the blue legend below, he calls "Forward". His reference frame is also shown (in blue of course).

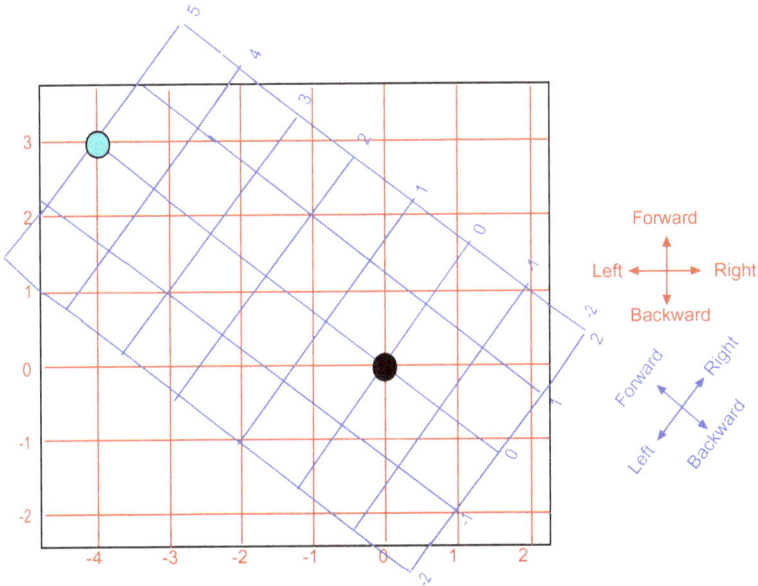

According to Bob's frame, the lunch is located zero yards to the left and 5 yards forward, at the point $(0, 5)$.

Now, I want to point out a few things that you might think are too obvious to mention — but in the next few chapters you'll see why they were worth mentioning anyway:

- *Alice and Bob completely agree about where the lunch is*, but *they assign different coordinates to that same location.* That is, they describe the same reality using different frames.
- There is no "one true frame". Alice's frame is no better and no worse than Bob's.
- Although any frame is as good as another, there is one frame that comes most naturally to Alice and a different frame that comes most naturally to Bob.
- The frame that comes most naturally to you depends on the direction you're facing.

- If she ever finds it convenient, Alice can always use Bob's frame to describe any location in the desert. But most of the time, she instinctively uses her own system. And likewise, of course, for Bob.

Before we leave the desert, there's one more thing to attend to, namely the coordinate transformations between Alice's frame and Bob's. If Alice assigns coordinates (x, y) to some point, and Bob assigns coordinates (x', y') to the same point, then these values are related by

$$x' = \frac{3}{5}x + \frac{4}{5}y \quad y' = -\frac{4}{5}x + \frac{3}{5}y \tag{4.1}$$

$$x = \frac{3}{5}x' - \frac{4}{5}y' \quad y = \frac{4}{5}x' + \frac{3}{5}y' \tag{4.2}$$

For example, Alice says that her lunch is under a tree with coordinates $(x, y) = (-4, 3)$. If you want to know the coordinates Bob assigns to that tree, you can use Equation (4.1) with $x = -4$ and $y = 3$. Plugging in those values, you'll find that $x' = 0$ and $y' = 5$, which, as we've already seen, are Bob's coordinates.

Going backward, if you plug $x' = 0$ and $y' = 5$ into Equation (4.2), you'll recover the fact that $x = -4$ and $y = 3$.

Exercise 4.1.   Alice has noticed an interesting little dandelion growing at the point she calls $(2, 1)$. Use Equation (4.1) to determine the coordinates Bob assigns to this point. Locate the dandelion on the graph above. Do the coordinates you just computed match what you see on Bob's grid? (Hints and answers for all exercises can be found in Appendix One.)

Exercise 4.2.   Bob has noticed a cactus growing at the point he calls $(-1, 3)$. Use Equation (4.2) to determine the coordinates Alice assigns to this point. Locate the cactus on the graph above. Do the coordinates you just computed match what you see on Alice's grid?

✳

> If you're not comfortable with trigonometry, it's safe to ignore this note.

You might wonder how I came up with Equations (4.1) and (4.2). The answer lies in the geometry. If you imagine rotating the desert counterclockwise around Alice, Alice's frame is transformed into Bob's frame. The general formula for a rotation is

$$x' = \cos(\theta)x + \sin(\theta)y \quad y' = -\sin(\theta)x + \cos(\theta)y \quad (4.3)$$

$$x = \cos(\theta)x' - \sin(\theta)y' \quad y = \sin(\theta)x' + \cos(\theta)y' \quad (4.4)$$

where θ is the angle of the rotation.

In this case, I chose to rotate the grid by an angle of $\theta = 53.13$ degrees, because it so happens that $\cos(53.13°) = 3/5$ and $\sin(53.13°) = 4/5$ are particularly easy numbers to work with.

Chapter 5

Invariants in the Desert

Here, once again, is the desert, with Alice and Bob's frames superimposed. Remember that both Alice and Bob are standing at the black dot and that Alice has left her lunch under a tree at the green dot:

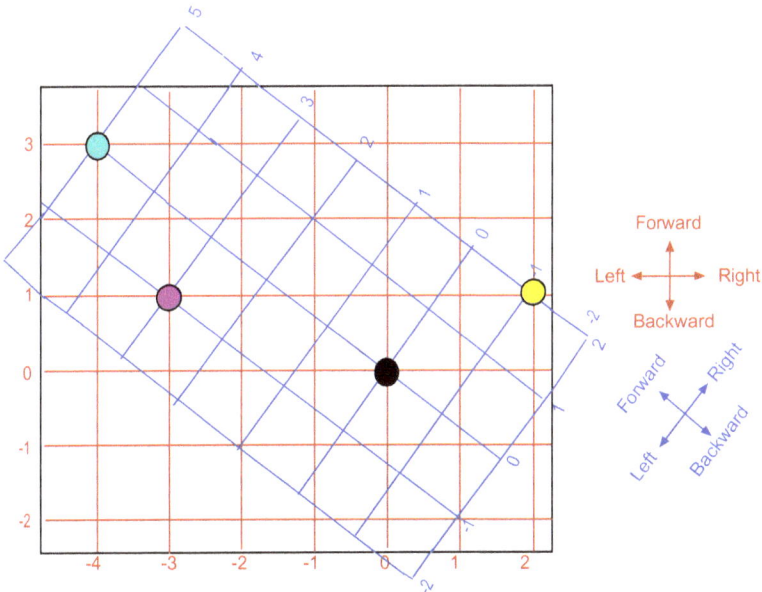

We've also added some more landmarks — a dandelion in yellow and a cactus in violet.

Now, suppose Alice wants to describe the location of the dandelion *relative* to the cactus. In her coordinates, the cactus is at $(-3, 1)$ and the dandelion is at $(2, 1)$. Relative to the cactus, the dandelion is 5 yards to the right and 0 yards forward. We record this information by writing

$$\Delta x = 5 \quad \Delta y = 0$$

For Bob, the cactus is at $(-1, 3)$ and the dandelion is at $(2, -1)$. So we write

$$\Delta x' = 3 \quad \Delta y' = -4$$

These coordinate *differences* (which we sometimes call *deltas*) are related to each other just as the coordinates themselves are. In other words, we can rewrite the coordinate transformations (4.1) and (4.2) in terms of differences and get

$$\Delta x' = \frac{3}{5}\Delta x + \frac{4}{5}\Delta y \quad \Delta y' = -\frac{4}{5}\Delta x + \frac{3}{5}\Delta y \qquad (5.1)$$

$$\Delta x = \frac{3}{5}\Delta x' - \frac{4}{5}\Delta y' \quad \Delta y = \frac{4}{5}\Delta x' + \frac{3}{5}\Delta y' \qquad (5.2)$$

✳ These equations continue to assume that Bob's frame is rotated 53.13 degrees from Alice's. If it were rotated by some other angle θ, we would use formulas derived from (4.3) and (4.4):

$$\Delta x' = \cos(\theta)\Delta x + \sin(\theta)\Delta y \quad \Delta y' = -\sin(\theta)\Delta x + \cos(\theta)\Delta y \qquad (5.3)$$

$$\Delta x = \cos(\theta)\Delta x' - \sin(\theta)\Delta y' \quad \Delta y = \sin(\theta)\Delta x' + \cos(\theta)\Delta y' \qquad (5.4)$$

Although Bob and Alice disagree about the leftward and forward distances from the dandelion to the cactus, they can at least agree on the overall as-the-crow-flies distance.

Let's check that.

Using the Pythagorean formula for distance, Alice computes

$$\sqrt{(\Delta x)^2 + (\Delta y)^2} = \sqrt{(5)^2 + 0^2} = 5$$

Meanwhile Bob computes

$$\sqrt{(\Delta x')^2 + (\Delta y')^2} = \sqrt{3^2 + (-4)^2} = 5$$

which is the same answer that Alice gets.

Exercise 5.1. Compute the distance from the (yellow) dandelion to the (green) tree using Alice's coordinates. Compute the same distance using Bob's coordinates. Make sure you get the same answer. Now, do the same for the distance from the tree to the (black) spot where Alice and Bob are standing.

Just as it was on Fifth Avenue, the distance between any two points is an *invariant* of the coordinate transformations, that is, it comes out the same no matter which frame you use.

There are two ways to see why this must be true. One is to grind through some algebra. The other is to trust the geometry. It's quite clear that rotating your frame — without moving either the tree or the cactus — cannot change the distance from the tree to the cactus.

<hr>

✳ For those who are interested, here is the algebra behind the observation that distance is an invariant:

Consider two objects, which Alice says are located at the points (x_1, y_1) and (x_2, y_2). To find the distance between these objects, she sets

$$\Delta x = x_2 - x_1$$
$$\Delta y = y_2 - y_1$$

and computes the distance as

$$d_A = \sqrt{(\Delta x)^2 + (\Delta y)^2}$$

Bob assigns (x'_1, y'_1) and (x'_2, y'_2) to the same objects, with the primed coordinates given explicitly by equation (4.1). He sets

$$\Delta x' = x'_2 - x'_1$$
$$\Delta y' = y'_2 - y'_1$$

and computes the distance as

$$d_B = \sqrt{(\Delta x')^2 + (\Delta y')^2}$$

(continued on next page)

(continued from previous page)

We can replace $\Delta x'$ and $\Delta y'$ with the formulas given in (4.1) to rewrite Bob's distance d_B as

$$d_B = \sqrt{\left(\frac{3}{5}\Delta x + \frac{4}{5}\Delta y\right)^2 + \left(-\frac{4}{5}\Delta x + \frac{3}{5}\Delta y\right)^2}$$
$$= \sqrt{(\Delta x)^2 + (\Delta y)^2} = d_A$$

This equality between d_B and d_A is exactly what we were out to check.

Our calculation assumes that Bob's frame is rotated 53.13 degrees from Alice's, so that Equation (5.1) applies. For an arbitrary rotation, we'd have used Equation (5.3) instead.

Exercise 5.2.   (This exercise is for those who have read the box directly above.)

Using Equation (5.3), check that the distance between any two points is an invariant of the coordinate transformations.

Finally, we've been assuming all along that Alice and Bob have agreed about which point to call the origin. But that doesn't have to be the case. If Bob is standing by the cactus instead of next to Alice, he might want to think of the (purple) cactus as the origin. That would change his frame from what you see in the left-hand panel following to what you see in the right-hand panel:

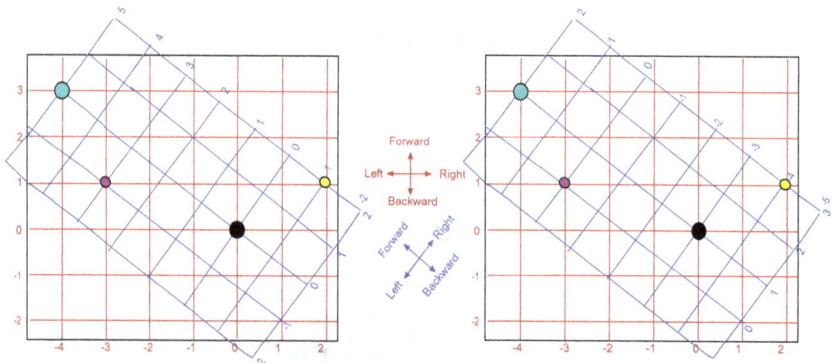

Here, Bob's *gridlines* are unchanged because he's still facing the same way, but the *labels* on those gridlines have been revised because he's standing in a new location. But none of his deltas have changed. He still says, for example, that to get from the cactus to the dandelion, you've got to go 3 yards right and 4 yards backward.

In a case like this, where Bob and Alice no longer agree on the location of the origin, you can no longer use the transformations (4.1) and (4.2) to go back and forth between Alice and Bob's coordinates. But you *can* still use the transformations (5.1) and (5.2) to go back and forth between the deltas — because the relocation of the origin has no effect on those deltas.

To summarize this chapter:

- When Bob and Alice agree on the location of the origin, you can use the transformations (4.1) and (4.2) to convert back and forth between Bob's and Alice's coordinates.
- Distances calculated using Bob's coordinates are always the same as distances calculated using Alice's coordinates. We express this by saying that these distances are *invariants* of the transformations.
- When Bob and Alice *don't* agree on the location of the origin, you can no longer use the transformations (4.1) and (4.2) to convert back and forth between their coordinates. But you *can* use the analogous transformations (5.1) and (5.2) to convert back and forth between their deltas.

PART II
FRAMES IN SPACETIME

Chapter 6

Frames in Spacetime

Alice has returned to Seventh Avenue, where Jeter, who is facing her, asks for directions to this evening's concert at Carnegie Hall. Her answer: "It's five blocks to the left and three hours from now."

Of course "five blocks to the left" is perfectly accurate *in Alice's coordinate system*, but not in Jeter's. Nevertheless, the information is perfectly useful to him, because he's sensible enough to apply the coordinate transformation that converts "five blocks to the left" into "five blocks to the right".

Since we're talking about coordinates, we're talking about geometry, so we need to introduce a form of geometry that accounts for both space and time. That means we need some pictures. Let's start with these maps of Seventh Avenue at various times of day:

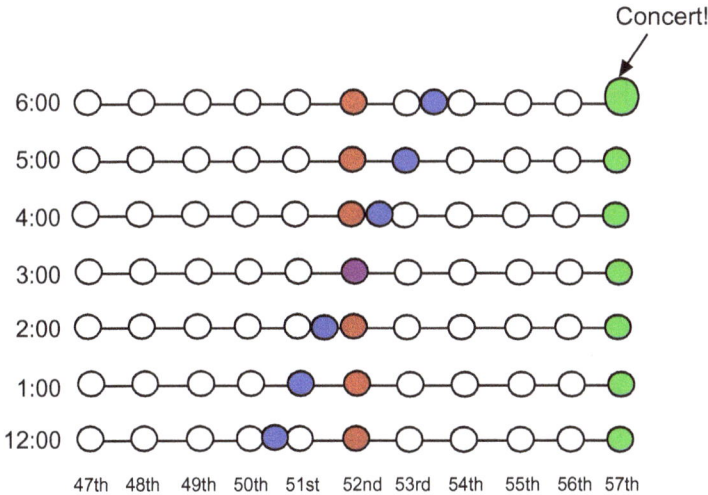

The red dots represent Alice, who spends the entire afternoon standing at the corner of 52nd Street, five blocks from Carnegie Hall at 57th. The blue dots represent Bob, who spends the afternoon skateboarding along Seventh Avenue at the leisurely pace of 1/2 block per hour. At noon, he's halfway between 50th and 51st. At 1:00, he's made it to 51st. At 3:00, he's made it to 52nd, right where Alice is (the purple dot indicates both Alice and Bob). At 6:00, he's halfway between 53rd and 54th.

The green dots represent Carnegie Hall, which is always at 57th street, and the big green dot at 6:00 represents the 6 PM concert at Carnegie Hall.

Of course it would be better to have a graph that can represent all the intermediate times between, say 2:00 and 3:00. That's easy enough:

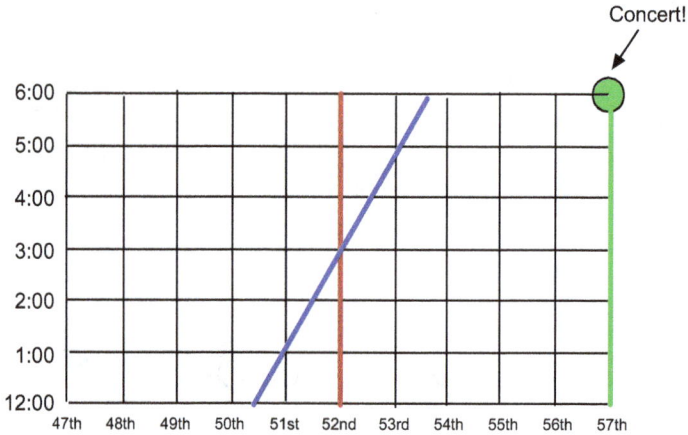

A map of Seventh Avenue is a one-dimensional space, where each point represents a location, such as Carnegie Hall. A map of *Seventh Avenue over time*, such as the map directly above, is a two-dimensional space (sometimes called a *spacetime*), where each point represents an *event*, such as tonight's concert.

On the spacetime map, landmarks and other objects — including Alice, Bob, or Carnegie Hall itself — are represented not by points but by *lines*, which are called their *worldlines*. The picture shows Alice's worldline in red, Bob's in blue, and Carnegie Hall's in green.

When Alice and Bob roam the two-dimensional desert, the corresponding spacetime is three-dimensional, and hence a bit difficult to depict on a flat piece of paper. When they pilot their spacecrafts through three spatial dimensions, the corresponding spacetime is four-dimensional (and hence extremely difficult to depict on a two-dimensional piece of paper or even a computer screen). For most of this book, we'll try to keep our pictures manageable by letting Alice and Bob move in only one dimension.

The grid lines on the map of spacetime — the lines showing 50th street, 51st street and so forth in one direction, and 2:00, 3:00 and so forth in the other — form a perfectly good frame. But we'd like to relabel these gridlines to form a slightly different frame, namely the one that is most natural for Alice-at-3:00 to employ. First, though, let's ask her to turn around and face *into* the page, just as you and I are, so that "left" and "right" mean the same things to Alice that they mean to you and me. Then her frame looks like this:

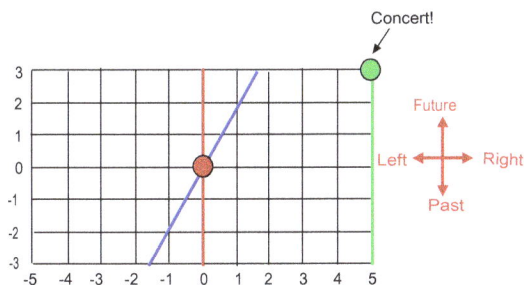

Alice-at-3:00 (represented by the red disk in the center) has decided to call her current time and position (that is, 3:00 and 52nd Street) "time zero" and "location zero". The concert is five blocks to the right and three hours in the future, at the point she now calls (5, 3).

Alice-at-4:00 (assuming she continues to hang out at 52nd Street) will have a new frame, with exactly the same gridlines, but with all the labels on the vertical axis shifted down by one hour.

Chapter 7

The First Principle of Relativity

Looking back at the previous graph, you can see that Bob's worldline is perfectly straight. This is because he's passing the cross-streets at a fixed rate of 1/2 block per hour. If that speed kept changing, his worldline would be steeper in some places and flatter in others, which is to say that it would not be a straight line at all.

Anybody whose worldline is straight is called an *inertial observer*. In our picture, Bob, Alice and Carnegie Hall are all inertial observers. Jeter, who has been hopping on and off his skateboard, performing wheelies, crashing into lampposts, and alternately speeding up and slowing down, is not.

There are exactly two key principles that drive all of relativity theory. Here is the first:

First Principle of Relativity: Alice — or any other inertial observer — has the right to consider herself stationary, and no experiment can prove her wrong. In fact, there is no "right" or "wrong" about it. There is no meaningful sense in which one inertial observer is more stationary than another.

Bob, for example, who according to Alice is skateboarding along Seventh Avenue toward Carnegie Hall, has every right to claim that he is standing still while Carnegie Hall moves leftward toward him.

(The principle does not apply to non-inertial observers. A skateboarder who keeps changing speed will feel inertial forces trying to push him off his skateboard, and so cannot plausibly claim to be stationary.)

Therefore, Bob considers every point on his own worldline to be at location zero. What about the rest of Bob's frame? You might expect it to look like this:

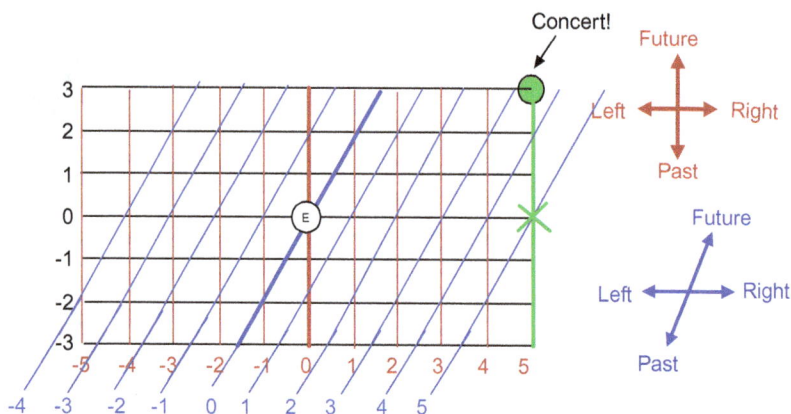

Let's take a little time to understand this graph.

The red vertical lines are part of Alice's frame. The blue slanted lines are part of Bob's frame. The black horizontal lines are part of *both* frames. Bob and Alice are both facing into the page, just as you are, so they agree on which way is right and which way is left.

Imagine yourself at the event labeled *E*, where Bob and Alice cross paths. They both label that event (0,0).

At this point, here's how Alice describes the world: "I am standing still. Bob is just passing me, skating rightward at half a block per hour. Carnegie Hall is standing perfectly still, five blocks to my right. The concert is five blocks to the right and three hours in the future."

Here's how Bob describes that same world: "I am standing still. Alice is just passing me, moving leftward at half a block per hour. Carnegie Hall is currently five blocks to my right, and is also moving leftward at half a block per hour. The concert is three-and-a-half blocks to the right and three hours in the future."

When Bob says "Carnegie Hall is currently five blocks to my right", he means that the event "Carnegie Hall right now" is located at the green *X*, five blocks to the right according to his coordinate system (and according to Alice's). (Note that the green *X* is on the blue coordinate line labeled 5.) But although Bob and Alice agree that Carnegie Hall is *currently* five blocks to the right, Bob does not agree with Alice's assertion that the *concert* is five blocks to the right. That's because, over the next three hours between now and the concert, Bob expects Carnegie Hall to move a block and a half closer. The concert (according to Bob) is at location 3.5, halfway between the blue coordinate lines labeled 3 and 4.

Who's right? Both, or neither, as you prefer to express it. The key thing is that neither is more right than the other.

They do, at least, agree on the time of the concert: it's three hours from now.

And that is relativity, as it was understood for hundreds of years before Albert Einstein. What we now know is that this picture is only an approximation to the truth. As long as everyone and everything is moving toward everyone and everything else at speeds that are small compared to the speed of light, the approximation is so good that we never notice it's not exact. At higher speeds, there's more to be said — and we are about to say it.

Chapter 8

The Second (and Last!) Principle of Relativity

Light travels at a speed of 670,680,000 miles per hour. That's pretty fast. Our friend Bob has been skating down Seventh Avenue at a speed of half-a-block per hour. That's a lot slower. At such slow speeds (and even at much much higher speeds) the preceding graph gives an extremely accurate, but still just approximate, picture of Bob's frame.

But when Bob revs his skateboard up to a much higher speed — say 335,340,000 miles per hour — the picture becomes starkly different.

The first thing we'll want to do is to choose units that will let us avoid throwing such large numbers around. A *light-hour* is the distance light travels in an hour; that is, it's 670,680,000 miles. From now on, instead of saying light travels at a speed of 670,680,000 miles per hour, we will say that it travels a speed of 1 light-hour per hour, or, more succinctly, that light travels at a speed of 1. Bob, on his 335,340,000-mile-per-hour skateboard, is then traveling at a speed of 1/2.

Now, let's transport Bob and Alice to some faraway version of Seventh Avenue, which is just like the version you already know about, except that each block is one light-hour long. Here is Alice, standing on a street corner. When her watch reads "time 0" — that is, at the event labeled E — she shines her flashlight rightward.

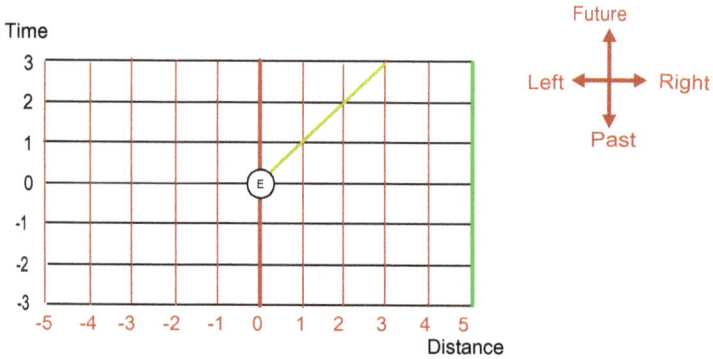

(All times are measured in hours and all distances in light-hours.)

The worldline of the light beam is shown in gold. It's important to realize that this light beam is running *parallel to the ground*, although the picture shows it slanting upward at a 45 degree angle. Remember that the upward direction on the graph represents *time*, not distance above the ground. You can see from the picture that after 1 hour, the light has traveled 1 light-hour, after 2 hours it has traveled 2 light-hours, and so on. The fact that light travels exactly 1 light-hour per hour is reflected in the 45-degree slant of its worldline.

Now let's add Bob to the picture. As he so likes to do, Bob is skateboarding to the right, now at half the speed of light (one-half light-hour per hour). He passes Alice just at the moment when she turns on her flashlight (at the circled event E), and notices that Alice is moving leftward. Here (in blue) are his worldline and (at least according to what we've said so far) his frame:

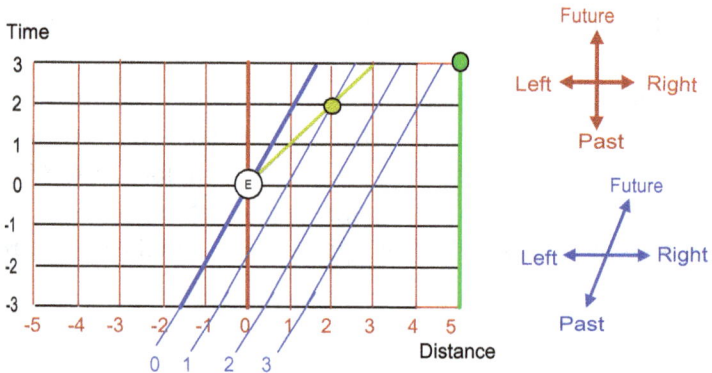

Now let's calculate the speed of the light beam, according to Bob. For this, we'll use Bob's frame. At time 0, the light beam is, according to Bob, at location 0. At time 2, the light beam is, according to Bob, at location 1 (as indicated by the gold circle). So, according to Bob, it takes two hours for the light beam to travel one light-hour. It is traveling, then, at speed 1/2.

When you think about it, this makes all the sense in the world. Alice, who considers herself stationary, sees the light beam traveling rightward at speed 1 and Bob traveling rightward at speed 1/2. The light, then, should be getting away from Bob at speed 1/2. (Likewise, if she saw the light traveling at speed 60 and Bob traveling in the same direction at speed 40, she'd expect the light to get away from Bob at speed 20.) So, of course Bob, who considers himself stationary, says that the light is moving at speed 1/2.

Yes, that makes all the sense in the world. But it's wrong. It's wrong because it runs directly counter to the second principle of relativity.

Second Principle of Relativity: All inertial observers must agree that the speed of light is exactly one light-hour per hour.

This Second (and last!) Principle of Relativity is one of the best-established facts in all of science. It is confirmed by a vast body of experimental evidence. It also runs entirely counter to everyday experience, where observers in motion with respect to each other can never agree on the speed of, say, a baseball.

Since Bob's calculation violates the Second Principle of Relativity, it must be the wrong calculation — which means we must have drawn his frame incorrectly. When we find his *true* frame, all of the pieces of relativity theory will fall into place.

PART III
COMPARING FRAMES

Chapter 9

Changing Frames: The Lorentz Transformation

Once again, Alice is standing on a street corner, when she is passed by Bob, who she says is traveling rightward at speed v. (Up till now, we've assumed $v = 1/2$.) Bob, of course, describes the situation a bit differently: He says that *he* is standing still, while Alice is traveling leftward at speed v. (Or perhaps he says she is traveling *rightward* at speed *minus v*, which means the same thing.)

They both assign coordinates $(0,0)$ to the event E where they pass each other. Alice (using her own frame of course) says that the Carnegie Hall concert occurs at location $x = 5$ and time $t = 3$. Bob, on the other hand, says that the concert occurs at a different place and time, because he uses a different frame.

In Chapter 7, we took a guess at Bob's frame. (That guess is illustrated in the figure at the top of page 32). In Chapter 8, starting near the bottom of page 36, we discovered that if that guess were right, Bob would compute the wrong speed of light.

So, our guess didn't work, and we still need to discover Bob's actual frame. We also need to discover the coordinate transformations — the formulas (analogous to Equations (4.1) and (4.2)) that will let us plug in Alice's coordinates x and t to compute Bob's coordinates x' and t'.

We will complete those tasks in the opposite order. First, here are the coordinate transformations:

$$x' = \frac{x - vt}{\sqrt{1 - v^2}} \tag{9.1}$$

$$t' = \frac{t - vx}{\sqrt{1 - v^2}} \tag{9.2}$$

All of the content of relativity theory is contained in these two formulas. Taken together, they are called the *Lorentz transformations*. Sometimes, we think of the two equations as a single package, so that the pair of equations is simply called the *Lorentz transformation* (without an *s* at the end).

Where ever did these formulas come from? How did anyone ever think of them? The answer is in Appendix II near the end of this book. (That material is somewhat more mathematically demanding than the rest of the book.) If you don't like taking things for granted, you can go read that appendix now, and then come back. Or, if you're willing to trust me, at least provisionally, you can stay right here and explore the consequences of the formulas and learn that, once you get used to them, they make all the sense in the world. Then, once you've learned to work with the formulas and seen how useful they are, you can turn to Appendix II to fill in the gap in your education.

Remember that Alice says the concert takes place at location $x = 5$ and time $t = 3$. Bob, who is traveling past Alice at velocity $1/2$, says that it takes place at some location x' and some time t'. Plugging $x = 5$, $t = 3$ and $v = 1/2$ into our formulas gives

$$x' = \frac{5 - 3/2}{\sqrt{1 - 1/4}} \approx 4.04 \quad t' = \frac{3 - 5/2}{\sqrt{1 - 1/4}} \approx .577$$

Bob says the concert takes place at location $x' = 4.04$ and time $t = .577$.

It's not too surprising that Alice and Bob disagree about the concert's location. Alice, after all, believes that Carnegie Hall is standing still, whereas Bob believes that it's moving leftward down

Seventh Avenue, and will be a lot closer by the time the concert starts. What's considerably more surprising (at least if you're new at this) is that Alice and Bob also disagree about the *time* of the concert. Alice says it will take place 3 hours from now. Bob says it will take place in just .577 hours, which is to say roughly 35 minutes.

Who's right? They both are. Inertial observers in motion with respect to each other always have equally legitimate ways of assigning times and locations to distant events. How is this possible? Again, the full answer to any question is embedded in Equations (9.1) and (9.2). All we have to do is unravel it.

We'll start by drawing Alice's frame, which is just the same as it was at the top of page 36, though we've made a few slight tweaks to the picture. First, we've zoomed in on the upper right quadrant. Next, we've changed the color of the horizontal grid lines from black to red. And finally, we've changed the label on the horizontal axis from "Distance" to "Space" (either of which is an abbreviation for "Distance in Space".) But all of those changes are cosmetic; Alice's frame is the same as it ever was. Her worldline is still the bold vertical line at $x = 0$.

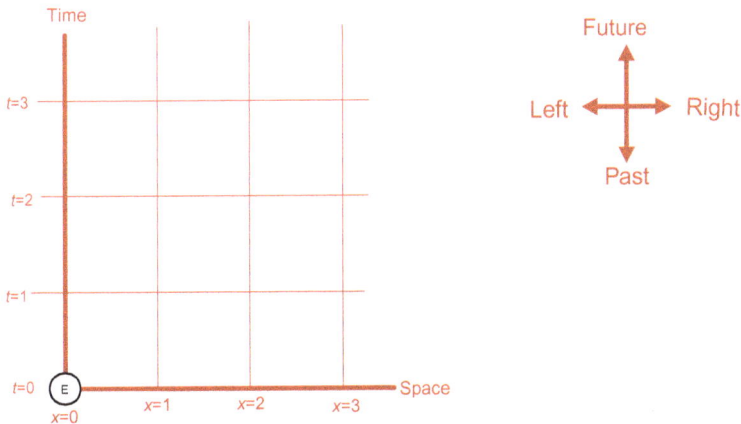

Now let's add Bob's worldline to the picture. Remember that Alice says Bob is traveling rightward at velocity v, so that (according to Alice) at time t, he has reached the location $t = vx$. His worldline consists of all the points that satisfy that equation. We've added it to the picture below (which is drawn as if $v = 1/2$):

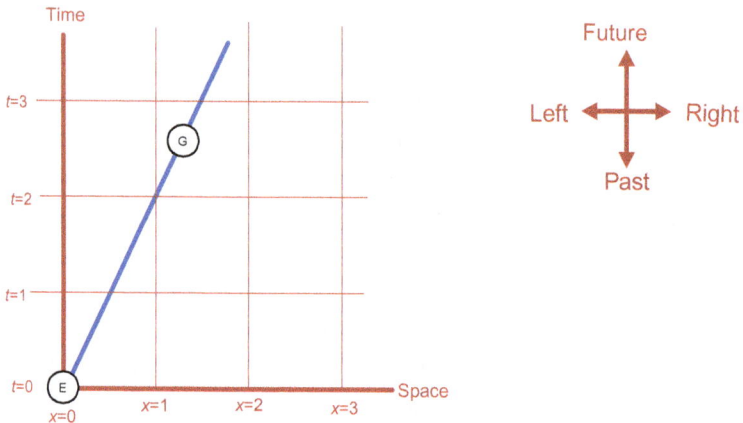

To see how the Lorentz transformation works, consider point G, which happens to be on Bob's worldline. Alice assigns some coordinates (x, t) to this point. (It looks like x is a little bit more than 1 and t is somewhere between 2 and 3.) Because G is on Bob's worldline, we know that $x = vt$.

We can apply the Equations (9.1) and (9.2) to discover the coordinates that Bob assigns to G:

$$x' = \frac{x - vt}{\sqrt{1 - v^2}} = \frac{vt - vt}{\sqrt{1 - v^2}} = 0$$

$$t' = \frac{t - xv}{\sqrt{1 - v^2}} = \frac{t - tv^2}{\sqrt{1 - v^2}} = \frac{t(1 - v^2)}{\sqrt{1 - v^2}} = t\sqrt{1 - v^2}$$

It should come as no surprise that $x' = 0$. After all, in Bob's frame, he is standing still, so every event on his worldline takes place at $x' = 0$. In other words:

> **The equation of Bob's worldline is $x' = 0$.**
> **Therefore his worldline is his time axis (just as Alice's worldline is *her* time axis).**

Exercise 9.1. (a) Starting with the coordinates that Bob assigns to event G (that is, $x' = 0$ and $t' = t\sqrt{1-v^2}$), use the Lorentz transformations to find the coordinates that Alice assigns to event G. (Of course we already know that Alice assigns the coordinates x and t, but the goal here is to check that the Lorentz transformations give the right answer.)

(b) Suppose H is an event on Alice's worldline, to which Bob assigns the coordinates x' and t'. Use the Lorentz transformations to find the coordinates that Alice assigns to event H.

Warning: Bob says that Alice is traveling *leftward*, with velocity $-v$ (note the minus sign). So, when you're translating from Bob's coordinates to Alice's, you'll need to use $-v$ instead of v in Equations (9.1) and (9.2).

Exercise 9.2. The graph near the top of page 44 reflects Alice's viewpoint in the sense that it depicts her time and space axes as vertical and horizontal. Redraw the same picture as seen from Bob's viewpoint.

Chapter 10

The Relativity of Simultaneity

Here again is the picture from page 44, showing Alice's frame and Bob's time axis:

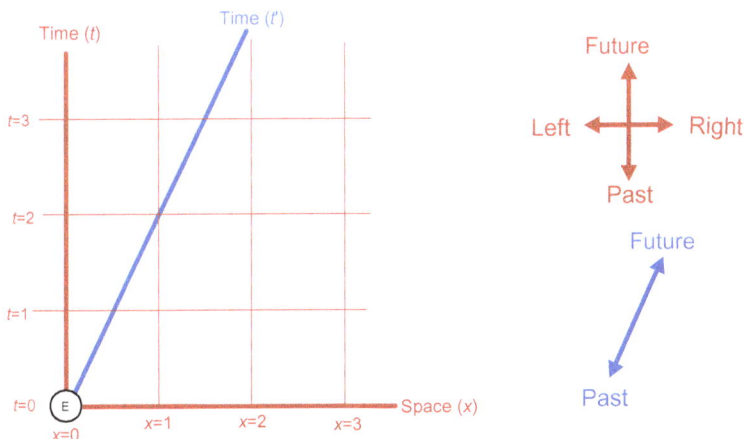

To complete the picture, we need Bob's space axis, given by the equation $t' = 0$.

The Lorentz transformation (9.2) says that $t' = (t - xv)/\sqrt{1 - v^2}$. This in turn tells us that $t' = 0$ exactly when $x = t/v$. Therefore,

> **In Alice's frame, Bob's space axis has the equation $x = t/v$.**

47

Here is the picture again, with Bob's space axis added:

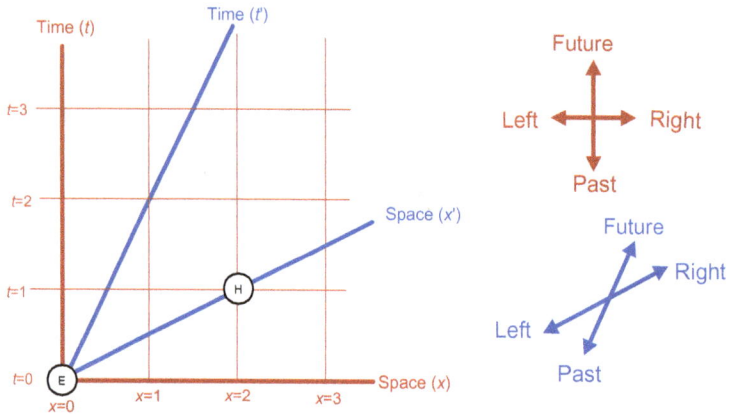

As before, we are drawing the picture as if $v = 1/2$.

As it happens, there's a car crash nearby, represented by point H, which you can see occurs at time $t = 1$ according to Alice, but at time $t' = 0$ according to Bob. When they tell their friends about it, Alice will say "First I met Bob, whizzing by me on a street corner, and then an hour later there was a car crash two blocks away." But Bob will say "That car crash happened at exactly the same moment when Alice and I met."

A reminder: We're measuring time in hours and light in blocks. In order to keep things simple, we want the speed of light to be one block per hour, so our blocks must be one light-hour long.

Do Bob and Alice contradict each other? No more than they'd contradict each other if they were facing different directions in the desert, assigning different meanings to the words "forward", "backward", "left" and "right". Each has his or her own, perfectly legitimate way of describing exactly the same events. Alice can acknowledge the legitimacy of Bob's description, while still insisting that only her own description matches her personal experience of the world, and Bob can say the same in reverse.

Note that at the moment when Bob passes Alice, Bob does not say "Oh my goodness, I just heard a car crash." That's because news of the crash takes time to travel from the crash site to Bob. That news might travel in any number of ways — someone could come running down the street yelling "Car crash, car crash!". Or Bob could actually *see* the car crash, which would require a light beam to travel from the crash site to Bob's eyes. Or he could hear it, which would require a sound wave to travel from the crash site to Bob's ears. Any of these things takes time, so Bob cannot know about the crash until after he passes Alice. But once he gets the information and pieces it all together, he will deduce that the crash happened at time 0, whereas Alice will deduce that it happened at (say) time 1.

The time of the car crash in Bob's frame is *the time that Bob says the crash actually occurred*, not the time he became aware of it. Likewise, suppose that at 1PM, you look up at the sky and see a spot form on the sun. You learned in school that light from the sun takes eight minutes to reach your eyes, so you deduce that the spot really formed at 12:52PM. In your frame, that spot formation occurred at 12:52PM, although you didn't become aware of it until eight minutes later.

Exercise 10.1. Suppose Bob's velocity is $1/2$, and the car crash takes place (according to Alice) at time $t = 1$, location $x = 2$. At what location does the car crash take place according to Bob?

Chapter 11

Maps of the World

The picture near the top of page 48 shows Bob's time and space axes. All that remains is to add some of his gridlines, which are parallel to the axes:

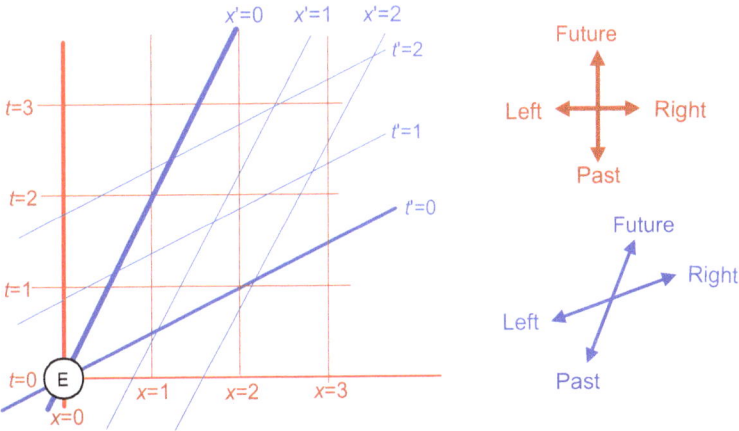

The red line labeled $x = 0$ is Alice's worldline; these are all the events that happen, according to Alice, at the same location as the event E. You can think of this line as showing "a history of everything that happens at location 0", according to Alice.

The blue line labeled $x' = 0$ is Bob's worldline; these are all the events that happen, according to Bob, at the same location as the event E. You can think of this line as showing "a history of everything that happens at location 0", according to Bob.

The red line labeled $t = 0$ shows all the events that, according to Alice, are simultaneous with the event E. You can think of this line as "a map of the world at time 0", according to Alice.

The blue line labeled $t' = 0$ shows all the events that, according to Bob, are simultaneous with the event E. You can think of this line as "a map of the world at time 0", according to Bob.

According to Alice, all of the events that occur along any one of the horizontal lines $t = 1$, $t = 2$ and so forth are simultaneous with each other. Each of these lines is called a *line of simultaneity* for Alice. You can think of these lines as maps of the world at various times according to Alice.

According to Bob, all of the events that occur along any one of the lines $t' = 1$, $t' = 2$ and so forth are simultaneous with each other. Each of these lines is a line of simultaneity for Bob. You can think of these lines as maps of the world at various times according to Bob.

For a reality check, let's see what the Lorentz transformation tells us about the speed of light, as measured by both Bob and Alice.

Just as Bob passes Alice at event E, Alice shines her flashlight rightward. Three hours later (according to Alice), the beam hits a lamppost three light-hours to the right (again, according to Alice). The event G where the beam hits the lamppost has, according to Alice, coordinates $x = 3$, $t = 3$:

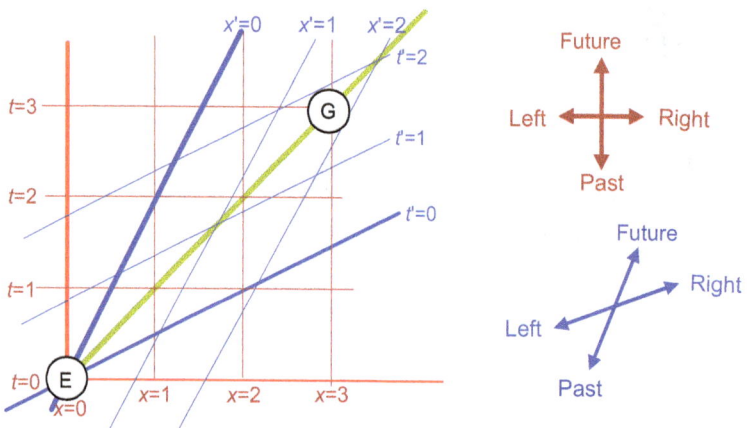

Applying the Lorentz transformation (9.1) and (9.2), we see that Bob says event G happens at

$$x' = (3 - 3v)/\sqrt{1 - v^2} \quad t' = (3 - 3v)/\sqrt{1 - v^2}$$

If $v = 1/2$, this gives

$$x' \approx 1.73 \quad t' \approx 1.73$$

which you can confirm is consistent with the blue grid lines on the graph.

Bob disagrees with Alice about many things. She says that lamppost is 3 light-hours away; he says it's 1.73 light-hours. She says the lightbeam takes 3 hours to reach the lamppost; he says it takes 1.73 hours. But they agree on one thing: They both calculate the speed of light as distance over time. Alice gets $3/3 = 1$ light-hour per hour, and Bob gets $1.73/1.73$, which is also 1 light-hour per hour.

The Lorentz Transformation, then, is fully compatible with the Second Principle of Relativity, which says that all inertial observers must agree that the speed of light is 1 light-hour per hour. This is further evidence that we've got the right equations!

Chapter 12

Spacetime and the Desert

We've just seen that Bob and Alice can assign different times to the same event. In this, they are no different from two observers standing in the desert, facing different directions, assigning different coordinates to the same tree — one saying that it's three yards forward and four yards to the left, while the other says it's five yards forward and zero yards to the left.

Indeed, when Bob is in uniform motion with respect to Alice, they are *facing different directions in spacetime*, as indicated by the slopes of their worldlines. Observers who face different directions will have different frames.

Alice and Bob can still communicate, by using the Lorentz transformations to figure out each others' coordinates. But Alice's coordinates are the ones she comes by naturally when she measures times and distances with her own wristwatch and her own yardstick. Ditto, of course, for Bob.

The same must true if Alice and Bob use any other sorts of measuring devices. If Bob's heart ordinarily beats 60 times per minute — keeping pace with the ticks on his wristwatch — then it must continue keeping pace with those ticks when he's in motion. (This is because of the First Principle of Relativity, which says that everything works the same way for Bob when he's moving with respect to Alice as when he's not.) So, if Alice says that Bob's wristwatch has slowed down, she must say his heart has slowed down as well — along with all his other bodily processes and the rate at which he ages. Therefore, when Alice and Bob disagree about the time associated with a given event, they also disagree about *their own ages* at the time of that event. We'll see some of the consequences of this in the next few chapters.

In Chapters 3 and 5, where Bob and Alice were standing on Seventh Avenue or in the desert, we discovered that the *distance* between any two objects was an *invariant* of the coordinate transformations — that's a fancy way to say that if you want to know how far it is from the dandelion to the cactus, Bob will compute one way and Alice will compute another, but they'll still get the same answer. In Chapter 11, after we'd brought time into the picture, we discovered that *the speed of light* is an invariant of the Lorentz coordinate transformations. When Alice shines her flashlight at a lamppost three blocks away, Alice and Bob each have their own ideas about how far the lamppost is and how long the light takes to get there, but when they calculate the speed of light, they get exactly the same answer.

Another thing we discovered back in the desert of Chapter 5 is that if Bob and Alice disagree about which point they've chosen to call the origin, we can no longer use the same simple formulas to convert Alice's coordinates to Bob's and back. But we *can* use essentially the same formulas to convert back and forth between their deltas. The same is true for the Lorentz transformations: When Alice and Bob choose different points to call the origin, the transformations (9.1) and (9.2) are replaced by

$$\Delta x' = \frac{\Delta x - v\Delta t}{\sqrt{1 - v^2}} \tag{12.1}$$

$$\Delta t' = \frac{\Delta t - v\Delta x}{\sqrt{1 - v^2}} \tag{12.2}$$

In the next few chapters, we'll work through some examples to flesh out the significance of these transformations.

PART IV

UNDERSTANDING TIME: A JOURNEY AROUND THE STARS

Chapter 13

A Journey to the Stars

Have I mentioned that Alice and Bob are twins? Bob, on their 20th birthday, has decided to migrate to a small planet revolving around Barnard's Star in the constellation Ophiuchus, 6 light-years away. That's pretty far, but at least he's got a fast ship that can travel at 3/5 lightspeed. Still, Alice is unimpressed. "Even at 3/5 lightspeed, that's a 10 year trip! By the time you get there, you'll be 30 years old!"

The following picture shows what Alice is thinking. The vertical red line is Alice's worldline (and the earth's worldline — at the scales we're talking about, the earth is a dot, and Alice is a dot on that dot, so we treat them as the same thing). The black line is the worldline of Barnard's Star, which sits at $x = 6$ light-years from earth. Bob's planned itinerary is the blue line. He leaves earth (event E) on their twentieth birthday, which they've both agreed to call "time zero." He arrives at his destination (event F) at time $t = 10$.

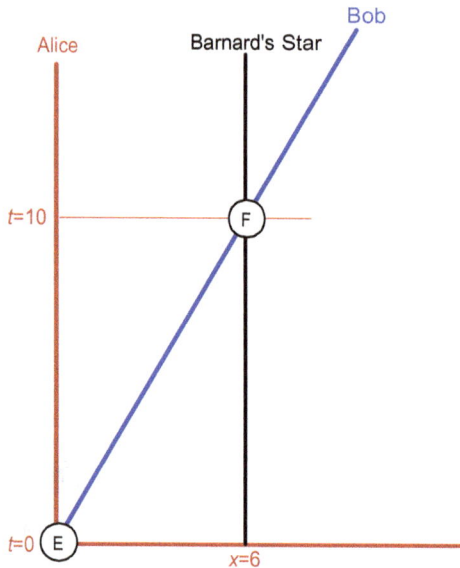

But of course, Bob (whose journey is now underway) assigns very different coordinates to the event F. Plugging $v = 3/5$, $x = 6$, and $t = 10$ into the Lorentz transformations ((9.1) and (9.2)), we get these equations for the coordinates of event F:

$$x' = 0$$

$$t' = \frac{10 - (3/5)6}{\sqrt{1 - (3/5)^2}} = 8$$

The first of these equations reminds us that throughout his trip, Bob views himself as stationary, with Barnard's Star traveling *toward* him. When it arrives, he'll still be at location 0, just as he always was.

The second is more interesting — it says that, according to Bob, he lands at Barnard's Star at time $t' = 8$ — on his 28th birthday.

What does Alice make of all this? Once she sees the pictures of 28-year-old Bob waving from Barnard's Star, she must conclude that all of Bob's timekeeping devices — including his own aging body! — have run slowly. According to Alice, Bob spent 10 years in space — but he aged only 8 years in the process.

So: *When Bob is in motion with respect to Alice, Alice sees all of his clocks — including his biological clocks — running slow.* In this particular case, Alice says that all of Bob's clocks have been running at 80% of full speed.

This is what the equations tell us, and therefore we must believe it. But here's what might bother you: Bob has every right to claim that *he* has been stationary all throughout his trip, while *Alice* was moving away *from him.* So, if relative motion causes Alice to say that Bob ages slowly, it should also cause Bob to say that Alice ages slowly.

And indeed that's the case. We can understand this better by adding a few grid lines to the diagram:

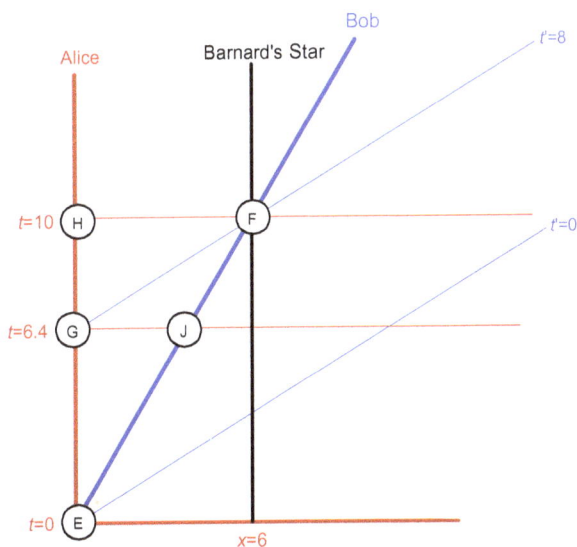

The two non-boldfaced blue lines are lines of simultaneity for Bob. As we've noted before, his lines of simultaneity have slope equal to his velocity, which in this case is $3/5$.

One line passes through the origin, which Bob has agreed to call $(x' = 0, t' = 0)$, so (because a line of simultaneity is, by definition, a line along which t' is constant), we must have $t' = 0$ all along this line.

The other line passes through F, to which we know Bob has assigned the time $t' = 8$, so t' must equal 8 all along this line.

Now take a closer look at the $t' = 8$ line. That line passes through $(x = 6,\ t = 10)$ and it has slope 3/5, so it must hit the axis at $(x = 0,\ t = 6.4)$. (Do the arithmetic!). That is the event we've labeled G, where Alice's watch strikes 6.4. So, Bob says "Here I am, just landing at Barnard's star, 8 years after I left. But *right at this very moment*, Alice's calendar-watch is striking 6.4 years, and her age is 26.4. Her watch, then — along with her heartbeat, her aging process, and everything else — has been running at 80% of full speed this entire time."

Of course, Alice quite disagrees. If you approach Alice when her clock is striking 6.4, she'll say Bob *right at this very moment* is at event J, only partway along his journey. (In fact, you can use the equation of Bob's worldline ($x = vt$) to check that at J, according to Alice, Bob has traveled only $x = 3.84$ light-years.

To sum up:

Alice says these things at different times:

- Event E: Today Bob and I are 20 years old and he is leaving on his journey.
- Event G: Today I am 26.4 years old and Bob is at event J, 3.84 miles from home. He ages at 80% speed, so while I have aged 6.4 years, he's aged only 5.12 and is now 25.12 years old.
- Event H: Today I am 30 years old and Bob is just arriving at Barnard's Star, 6 light-years away. He ages at 80% speed, so he is currently 28 years old.

Bob says these things at different times:

- Event E: Today Alice and I are 20 years old and I am leaving on my journey — though I prefer to say that I am staying still, while Alice moves away from me and Barnard's Star moves toward me.
- Event F: Today I am 28 years old, and am landing at Barnard's Star. Alice ages at 80% speed, so she is currently 26.4 years old.

Exercise 13.1. When Bob is at event J, how old does he say he is? How old does he say Alice is?

Exercise 13.2. Instead of supposing that Bob is traveling at velocity 3/5 with respect to Alice, suppose that he's traveling at some other velocity v. At what fraction of their normal speed does Alice say that Bob's clocks are running? (Answer: $\sqrt{1 - v^2}$.)

Chapter 14

A Journey from the Stars

In the preceding chapter, Bob made a one-way journey from Earth (where Alice lives) to Barnard's Star. Therefore, both Bob and Alice were present at the event E that marked the start of Bob's journey.

In this chapter, we'll tell a different story, where Bob makes a one-way journey from Barnard's Star to Earth. Therefore, both Bob and Alice will be present at the event that marks the *end* of Bob's journey.

So, we assume now that Bob starts out on Barnard's Star and travels to earth at $3/5$ lightspeed (though we now have to call his velocity $-3/5$, not $+3/5$, because he'll be traveling leftward). The bold blue line shows his itinerary, with departure at event K and arrival at event L:

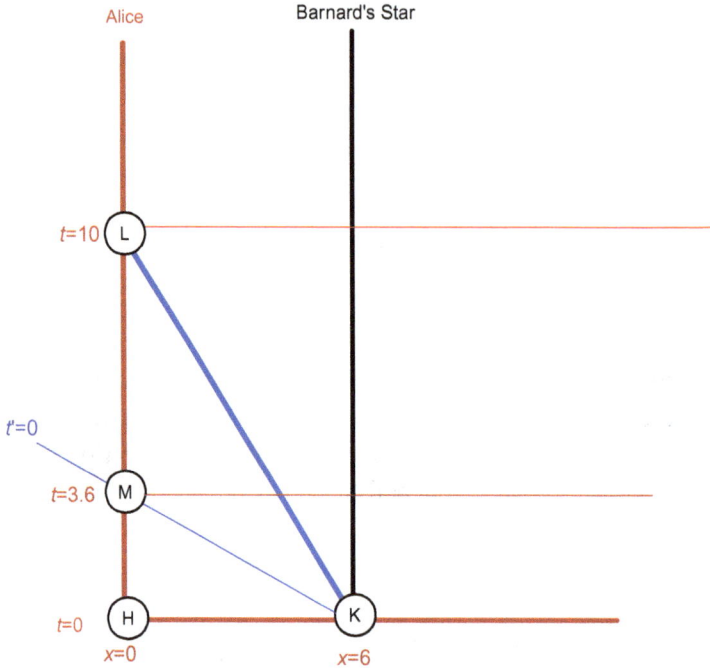

Prior to Bob's departure, he is chilling out at Barnard's Star, at rest with respect to Alice. Therefore, his coordinate lines are the same as Alice's, and they both agree that events H and K are simultaneous. They can therefore agree to say that both these events happen at time 0.

But the instant Bob takes off, his frame changes. He no longer considers H and K simultaneous. Now his line of simultaneity (with slope $-3/5$) runs from K through M, where Alice's clock strikes 3.6 years. So:

- The instant before Bob takes off, he says "Right now, Alice's clock reads 0."
- The instant after Bob takes off, he says "Right now, Alice's clock reads 3.6."

Bob switches in an instant from saying "Alice's clock now reads 0" to saying "Alice's clock now reads 3.6". This isn't because of any odd behavior on the part of Alice's clock. It's entirely because Bob (at event K) changes frames, from a frame in which K is simultaneous with H to a frame in which K is simultaneous with M.

In Chapter 1 we met Jeter, who thought it was *crazy* when Bob, by turning around, seemed to make Carnegie Hall jump all the way from 5-blocks-right to 5-blocks-left. Many students of relativity think (at first) that it's *crazy* when Bob, by accelerating his ship, seems to make Alice's clock jump 3.6 years forward. If that's how you feel, you're in excellent company — but just like Jeter, you're going to have to get over it.

Eventually, Bob arrives on earth at event L, where Alice's clock is striking 10. What does Bob's clock say at that point?

We can't simply use the Lorentz Transformations in the form of (9.1) and (9.2), because these apply only when Bob and Alice have agreed on a common origin. So, we use the Delta-form of the Lorentz Transformations, given by (12.1) and (12.2). The Delta from K to L, according to Alice, is $(\Delta x = -6, \Delta t = 10)$. Applying the transformation (12.2) we find that

$$\Delta t' = \frac{\Delta t - v\Delta x}{\sqrt{1 - v^2}} = \frac{10 - (-3/5)(-6)}{4/5} = 8$$

That is, Bob says events K and L are separated by 8 years. That's how long he says his journey takes.

Here, then, are Bob's and Alice's accounts of the journey:

Alice says:

- When Bob took off, his clock said 0.
- When Bob landed, my clock said 10 and his clock said 8.
- Therefore throughout the journey, his clock must have run at 80% speed.

Bob says:

- When I was about to take off, Alice's clock said 0.
- The instant I took off, her clock jumped ahead to 3.6.
- When I arrived, my clock said 8 and hers said 10 (so while my clock had advanced by 8 years, hers had advanced by only 6.4 years).
- Therefore throughout my journey, her clock must have run at 80% speed.

Note the symmetry. Alice says Bob's clocks run slow. Bob says Alice's clocks run slow by exactly the same amount. The one key difference between them is that Bob, unlike Alice, switched from one frame to another when he took off. Yet this one difference has remarkable consequences, as we'll see in the next chapter, where we combine Bob's two one-way trips into a single round-trip journey.

Chapter 15

A Round-Trip Journey: How Travel Keeps You Young

Now let's change the story a little: As before, Alice and Bob are twins who live on earth. On their 20th birthday, Bob decides to make a *round-trip* journey to Barnard's Star, 6 light-years from earth. He and Alice synchronize their calendar-watches to zero, then he takes off at 3/5 light speed, reaches Barnard's Star, turns around instantly, and returns at the same speed.

Here is Alice's (vertical) worldline in boldface red, Bob's outbound worldline in boldface blue, and Bob's inbound worldline in boldface green:

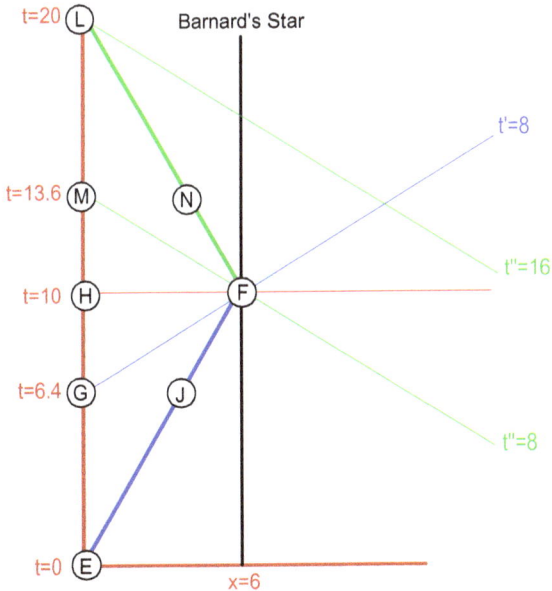

The outbound part of the journey is just as in Chapter 13 (illustrated on page 61). Event E is Bob's departure at time 0 on everyone's watch. Event F is Bob's arrival (and quick turn-around) at time 8 according to his watch and time 10 according to Alice's.

The return journey is just as in Chapter 14 (illustrated on page 66), where we computed that it also takes 8 years according to Bob. So, if he leaves the star when his watch says $t' = 8$, he must arrive home (at event L) when his watch says $t' = 16$ — just as he's celebrating his 36th birthday and Alice is celebrating her 40th.

To be more precise, the return journey is just as in Chapter 14 *except for the departure and arrival times*. The bold blue line from Chapter 14 has been shifted upward by 10 hours (as measured on Alice's time axis) to become the bold green line in the current picture.

Note that the instant Bob turns his ship around, his frame changes from the blue to the green (just as your frame changes the instant you turn around in a desert). In that instant, the times he assigns to events on earth all change (just as the leftward and forward distances you assign to a cactus in the desert change). In the blue frame (before he turns around), he says that events F and G are simultaneous. In the green frame (after he turns around), he says that events F and M are simultaneous.

Here is how each of the twins accounts for their recollections of Bob's journey:

Alice says:

- Bob left on our 20th birthday, when both of our watches were set to zero.
- After 6.4 years, he was at event J, partway to Barnard's Star.
- After 10 years, he arrived at Barnard's Star, having aged only 8 years, and immediately turned his ship around.
- After 13.6 years, he was at event N, partway home.
- After 20 years, he arrived home, having aged only 16 years.

Bob says:

- I left on our 20th birthday, when we each set our watches to zero.
- After 8 years, I arrived at Barnard's Star. At that time, Alice had aged only 6.4 years and she was claiming I was still at event J, only partway there.
- On arrival, I immediately turned my craft around, whereupon Alice's age immediately jumped from 26.4 to 33.6, and she was claiming that I was at event N, already partway home.
- After 16 years, I arrived home, to find that Alice had aged 20 years.

Part of Bob's description is "I immediately turned my craft around, whereupon Alice's age immediately jumped from 26.4 to 33.6." As always, it's important to recognize that he's not claiming anything magical here. Alice's age jumps forward in exactly the same sense, and for exactly the same reason, that when you turn your body ninety degrees, Des Moines, Iowa can immediately jump from being 1000 miles ahead of you to being 1000 miles to your right. It's Bob's frame that suddenly changes, not anything about Alice.

Bob is using different frames to describe different parts of his journey. There's a (blue) frame for Outbound Bob and a (green) frame for Inbound Bob. He uses the Outbound frame to describe the first half of the journey and the Inbound frame to describe the second half. These are the descriptions that fit Bob's recollections.

Note that Bob claims both that Alice consistently aged more slowly than he did *and* that she was older than him at the end of the journey — all of which is possible because of the sudden jump in her age at the moment when he turned around (and therefore changed frames).

As in the earlier examples, involving 1-way trips, Bob and Alice use different frames to describe the same events. But in the current example, something more startling has occurred: Everyone must agree that Bob is the same age as Alice when he departs, and everyone must agree that he is 4 years younger than Alice when he returns. These are objective truths that have nothing to do with frames.

Exercise 15.1.	Describe Bob's entire round-trip journey in the frame of Outbound Bob. (Outbound Bob accompanies Bob on his outward journey and continues onward at the same speed in the same direction even after Bob himself turns around.)

Exercise 15.2.	Describe Bob's entire round-trip journey in the frame of Inbound Bob. (Inbound Bob has always been traveling toward Earth from beyond Barnard's Star, arrives at Barnard's Star just as Bob is leaving, and accompanies Bob on his return journey.)

Exercise 15.3.	Alice and Bob are located 1 light-year apart, and stationary with respect to each other. Their clocks are synchronized. When the clocks strike noon, they start moving toward each other at the same speed. Now Alice claims that she is stationary and Bob is moving, so Bob's clock runs slow. Likewise, Bob claims that he is stationary and Alice is moving, so Alice's clock runs slow. When they pass each other, their clocks agree. Draw a spacetime diagram to illustrate their worldlines. Using that diagram, explain (from Alice's viewpoint) why their clocks agree even though Bob's is running slow. Do the same with Bob and Alice reversed.

Exercise 15.4.   Alice and Bob are twins who live on earth with their Mom. At time 0 (according to all of their clocks), Alice takes off heading west and Bob takes off heading east, both traveling at 3/5 the speed of light. After 10 years (according to their Mom), they both instantly turn around and head home, still at 3/5 the speed of light. Another 10 years later (according to their Mom), they both fly by the earth, waving to each other as they pass.

(a) At the time when Alice and Bob wave to each other, how does Alice fill in the blanks in the following description of her journey.

"I left earth _ years ago, and traveled for _ years at speed _ , during which my clock moved at _ times normal speed, so that it advanced a total of _ years. Then I turned around. Since then it's been _ years, during which my clock moved at _ times normal speed, so that it advanced a total of another _ years. That's why my clock now says _ ."

(b) At the time when Alice and Bob wave to each other, how does Alice fill in the blanks in the following description of Bob's journey.

"Bob left earth _ years ago, and traveled for _ years at speed _ , during which his clock moved at _ times normal speed, so that it advanced a total of _ years. Then he turned around and traveled another _ years, during which his clock moved at _ times normal speed, so that it advanced a total of another _ years. That's why his clock now says _ ."

Chapter 16

A Birthday Party

At the moment when Bob reverses direction, he says that Alice's age jumps from 26.4 years to 33.6 years. What does that look like on Bob's video screen, where he's monitoring Alice's webcam?

Answer: It looks like nothing unusual whatsoever! Indeed, Alice's webcam is beaming a signal to Bob's monitor. His monitor displays that signal. Bob's turnaround can't affect the path of the signal, and so can't affect what's on his screen.

So if he sees nothing unusual, what makes Bob say that Alice's age has jumped?

To answer that question, let's think about exactly what Bob's screen is showing at the moment when he arrives at Barnard's Star and turns around. We'll let Alice figure this out for us. Here's what she says:

- Bob arrived at Barnard's Start at time $t = 10$.
- Barnard's Star is 6 light-years from earth.
- Electromagnetic signals travel at the speed of light, so he was receiving a signal sent 6 years earlier — that is, at $t = 4$.
- At $t = 4$, I was celebrating my 24th birthday (because, as you'll recall, we've defined Alice's 20th birthday to be $t = 0$).
- Therefore, at the moment Bob was turning around at Barnard's Star, he was seeing the livestream of my 24th birthday party.

The picture below shows (in gold) the worldline of Alice's webcam signal, leaving from Alice's 24th birthday party (at T) and arriving at Bob's video screen (at F).

When Bob receives the signal (at event F of course), he is not so naive as to think that Alice is partying *right now*, because he knows the signal has taken some time to reach him. In fact, the signal left from event T, where $x = 0$ and $t = 4$. Applying the Lorentz transformation, we see that according to Bob, that party took place at $x' = -3$ and $t' = 5$. (Check the arithmetic!) Here's how he reasons:

- Here I am, at event F. The current time is $t' = 8$.
- Alice's party took place at $t' = 5$, exactly 3 years ago. Therefore this signal took 3 years to reach me. That means Alice was 24 years old exactly 3 years ago.
- You might think that makes Alice 27 years old today, but you'd be forgetting that she ages at 80% speed. In 3 years, she'd have aged only 2.4 years. So actually she is 26.4 years old today. Alice at this moment is at event G.

And in fact the picture confirms Bob's reasoning: His $t' = 8$ line of simultaneity shows that events G and F are, in Bob's frame, simultaneous.

Now Bob takes off for earth, traveling along the green worldline. His new frame is indicated by the green lines of simultaneity below:

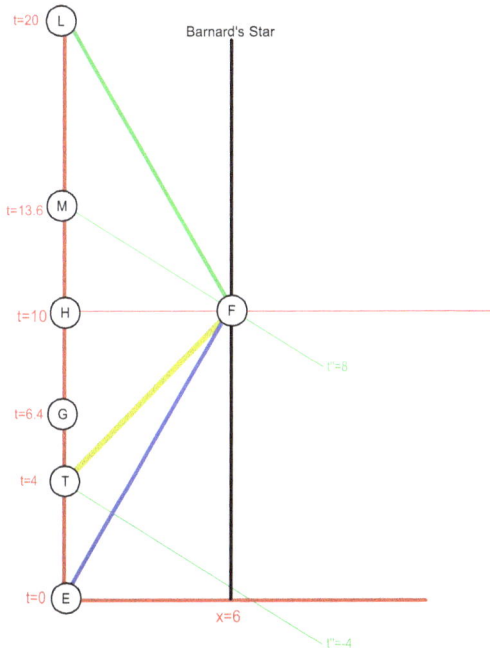

Alice's birthday party continues to display on Bob's video screen, with no interruption. But we'll soon see that Bob's *interpretation* of the video stream is now changed dramatically.

Let's first compute the coordinates of event T in Bob's new (green) frame. Because he and Alice no longer share an origin, we have to use the Delta form of the Lorentz transformation (equations (12.1) and (12.2).) From event F to event T we have $\Delta x = -6$ and $\Delta t = -6$, so $\Delta x' = \Delta t' = -12$. (Check the arithmetic, remembering that Bob's velocity is now $-3/5$.)

Here's how Bob now reasons:

- Here I am at Barnard's Star, 28 years old. I'll continue to call this time $t' = 8$.
- Alice's party took place at $\Delta t' = -12$, fully 12 years ago. Therefore this signal took 12 years to reach me. That means Alice was 24 years old exactly 12 years ago.

- You might think that makes Alice 36 years old today, but you'd be forgetting that she ages at 80% speed. In 12 years, she'd have aged only 9.6 years. That makes her 33.6 years old today. Alice at this moment is at event M.

And in fact the picture confirms Bob's reasoning: His $t' = 8$ line of simultaneity shows that events M and F are, in Bob's new frame, simultaneous.

If you stand in the center of the United States facing north, you'd report that California is to your left. If you turn south to face Mexico, you will report that California is now on your right. This doesn't mean you're repudiating anything you believed in the past; it means only that you now prefer to describe things in a very different language. And so it is with Bob: When he's facing one direction in spacetime, he describes "what Alice is doing right now" one way; when he's facing another direction, he describes it very differently.

We've said that at the instant Bob turns around, there is no change in the image on his video screen; that image is determined by the light that's striking his spaceship at the moment, and is unaffected by his velocity. On the other hand, the *speed* of the video on his screen will certainly change after he turns around, because he's now moving *toward* the video source whereas a moment ago he was moving away from it. We'll explore this a bit more in some of the exercises below.

All this might tempt you to believe that relativity is only about how Bob chooses to *describe* reality, as opposed to the workings of reality itself. But the fact remains that, based only on our analysis of these descriptions, we've proved that when Bob returns to earth, he's 4 years younger than Alice. If he'd traveled farther or faster, he could have come back 400 years younger than Alice — still in the prime of life but 400 years into Earth's future. We can *discover* such facts by contrasting Bob's descriptions with Alice's, but the facts we discover transcend those descriptions.

Exercise 16.1. Alice lives on earth. Bob lives on a planet 8 light-years away, which is stationary with respect to the earth. Their clocks are synchronized. At the moment when both clocks read zero, Alice sends a light-signal to Bob, and Bob takes off in his spaceship, traveling toward Earth at 4/5 lightspeed. When the signal arrives, what is the time on the ship's clock?

Exercise 16.2. Bob travels away from earth at speed 1/2. He and Alice, who stays behind, both set their clocks to zero at the moment when he leaves. When her clock says $t = 2$, Alice sends him a message that travels at the speed of light (that is, speed 1).

(a) According to Alice, when does the message arrive, and how far away was Bob when it got there?

(b) According to Alice, what did Bob's clock say at the moment when she sent the message? What did it say when the message arrived?

(c) According to Bob, where and when was the message sent? How long did it take to get to him? What time was it when the message arrived?

Exercise 16.3. Bob is traveling away from earth at some speed v, watching a movie that is being livestreamed from the earth. At what speed does the movie appear to run on Bob's screen?

Exercise 16.4. Repeat the previous exercise with Bob traveling *toward* earth at speed v.

Exercise 16.5. Repeat the previous exercise with Bob *orbiting* the earth at speed v.

Exercise 16.6. Alice is standing on a train platform, watching a boxcar travel west. At the east end of this boxcar stands Bob, with a lamp over his head. At the west end there is a mirror:

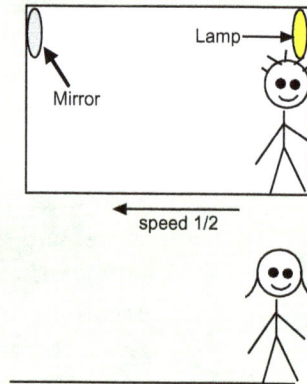

Just as the lamp passes Alice, it emits a single light beam, which hits the mirror and bounces back to the lamp.

(a) According to Bob, which takes longer, the light beam's journey from the lamp to the mirror, or its return journey from the mirror to the lamp? (**Hint:** This is a very easy question.) (Answer: According to Bob, the light travels the the length of the boxcar in each direction, so both journeys are equally long.

(continued on next page)

(continued from previous page)

(b) According to Alice, which takes longer, the light beam's journey from the lamp to the mirror, or its return journey from the mirror to the lamp? (Answer: According to Alice, the return journey is shorter, because the lamp is moving toward the mirror.)

(c) Draw a spacetime diagram to illustrate Alice's worldline, Bob's worldline, and the worldline of the light beam. Add Alice's and Bob's lines of simultaneity through the event where the light beam is emitted, the event where it bounces off the mirror, and the event where it returns to its starting point.

(d) Explain how your diagram illustrates the answers to parts (a) and (b).

Exercise 16.7. Alice and Charlotte are located one light-second apart and stationary with respect to each other. Bob is in motion at speed $3/5$, passing first Alice and then Charlotte. When he passes Alice, he turns on his stopwatch and Alice sends a light-signal to Charlotte. When Charlotte receives the signal, she turns on her stopwatch. When Bob passes Charlotte, they both turn their stopwatches off.

(a) How much time has Bob's stopwatch recorded?

(b) How much time has Charlotte's stopwatch recorded?

(continued on next page)

(continued from previous page)

(c) According to Charlotte, how long was Bob's stopwatch running? At what rate did it run?

(d) According to Bob, how long was Charlotte's stopwatch running? At what rate did it run?

Exercise 16.8.

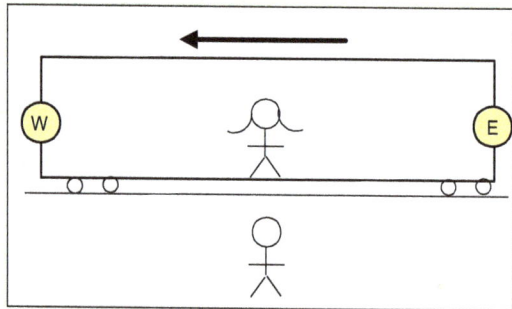

Alice rides in the center of a westbound train car with (yellow) clocks mounted on the east and west walls. The clocks are synchronized with her watch.

At 3PM by Alice's watch, she sends light-signals in both directions. Just as she sends the signals, she passes Bob, standing on the platform. His watch also says 3PM. The signals reach the clocks when they say 4PM, causing both clocks to stop.

Bob says that because the train is moving westward, the westward-moving signal has farther to travel, so Clock E must have been the first to stop.

(continued on next page)

(continued from previous page)

Alice says: "Don't be silly, Bob. You can see perfectly well for yourself that the clocks stopped at exactly the same time. They're both still showing 4PM."

(a) Use a spacetime diagram to illustrate how Bob and Alice can both be right.
(b) According to Bob, at what times did the clocks stop?
(c) According to Bob, how fast were the clocks running before they stopped?

PART V
THE ORDER OF EVENTS

Chapter 17

What We Can All Agree On:
The Spacetime Interval

There's a lot that Bob and Alice disagree about (or at least describe in very different languages). In this chapter, we'll talk about what they can agree on.

Here, once again, is a picture of Bob's journey, showing his departure from earth at E, his arrival at Barnard's Star and immediate turnaround at F, and his return to earth at L:

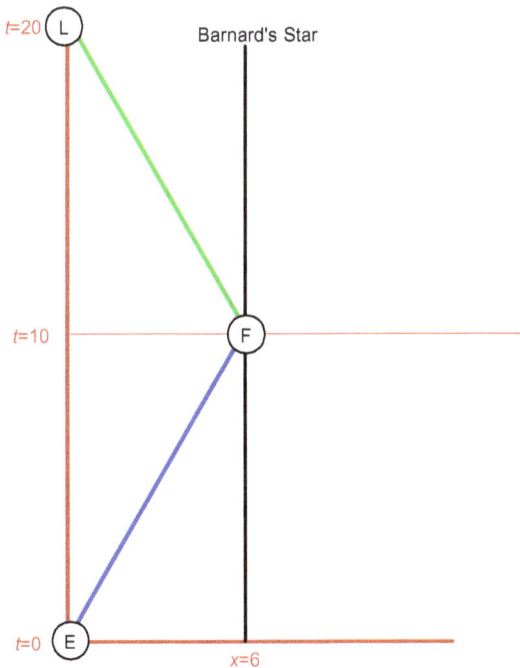

If we ask Alice "How far is it from event E to event F?", she'll answer that these events take place 6 light-years apart in *space* and 10 years apart in *time*. In brief, she says that

$$\Delta x = 6 \quad \Delta t = 10$$

If we ask Bob the same question, he'll say that events E and F take place 0 light-years apart in space and 8 years apart in time. So,

$$\Delta x' = 0 \quad \Delta t' = 8$$

This is the same sort of thing that happened to Alice and Bob when they were roaming the desert, back in Chapter 4 (page 15). They disagreed about the left–right distance from the cactus to the tree, and they disagreed about the forward–backward distance from the cactus to the tree. But in Chapter 5 (page 20), we saw that they could at least agree about the distance "as the crow flies". In other words, the "as the crow flies" distance is an invariant, though the left-right and forward-backward distances are not.

We'd like to find something analogous to the as-the-crow-flies distance that Alice and Bob can agree on. An obvious thing to try is to the analogue of the Pythagorean distance formula:

$$\sqrt{(\Delta t)^2 + (\Delta x)^2}$$

But using this formula, Alice concludes that events E and F are separated by the amount $\sqrt{10^2 + 6^2}$ while Bob gets $\sqrt{8^2 + 0^2}$ — not at all the same thing!

On the other hand, if we change the plus sign to a minus sign and define the *spacetime interval* between events E and F to be

$$\sqrt{(\Delta t)^2 - (\Delta x)^2} \tag{17.1}$$

then Alice computes the interval as

$$\sqrt{(\Delta t)^2 - (\Delta x)^2} = \sqrt{10^2 - 6^2} = \sqrt{64} = 8$$

while Bob computes the interval as

$$\sqrt{(\Delta t')^2 - (\Delta x')^2} = \sqrt{8^2 - 0^2} = 8$$

Now their calculations agree.

In fact, if *any* two observers calculate the spacetime interval between *any* two events, their calculations will agree. We express this by saying that *the spacetime interval between any two events is an invariant of the Lorentz transformations.*

Exercise 17.1. Check that the spacetime interval really is an invariant of the Lorentz transformations. In other words, suppose that Bob is traveling at some velocity v with respect to Alice, and that E and F are two events which Alice says differ by some amount Δx in space and some amount Δt in time. Use the Lorentz transformations to express the differences $\Delta x'$ and $\Delta t'$. Then compute the interval between E and F in Alice's coordinates and in Bob's coordinates, and check that the two calculations agree.

The spacetime interval is expressed in units of *distance*, so that the interval between E and F is 8 *light-years*. So, you should think of the interval as the *length* of the line segment connecting E and F (keeping in mind that in spacetime, the formula for length contains a minus sign, not a plus sign.)

You might have been troubled by the fact that $(\Delta t)^2$ is measured in units of *time squared*, while $(\Delta x)^2$ is measured in units of *distance squared*, so that it appears we have no business subtracting one from the other. In fact, we've been a little sloppy here. The precise formula for the spacetime interval is

$$\sqrt{(c\Delta t)^2 - (\Delta x)^2}$$

where c is the speed of light, measured in units of (distance over time). The factor of c makes the units agree on both sides of the minus sign. However, in our chosen system of units, c is numerically equal to 1, so we've been omitting it from our formula.

(continued on next page)

(continued from previous page)

> If we'd chosen some other system of units (such as *hours* for time
> and *miles* for distance) then it would be important to retain the
> *c* in the formula.

Refer again to the picture on page 87. During the course of Bob's
journey from E to F, his clock advances by 8 years. That's because
a clock is a spacetime odometer. If you carry a clock with you on
any straight-line journey, the passage of time that it records is equal
to the *length* of your journey in spacetime — where we interpret the
length of a line segment to mean the spacetime interval from one
endpoint to the other.

For example, in the diagram near the start of this chapter, the
length of the blue line is defined to be the interval from E to F, which
we've just calculated is 8. You can easily calculate that the length
of the green line — that is, the interval from F to L — is also 8.
So, Bob's total journey from E to L has length 16, and his clock
shows that the journey takes 16 years. But Alice's journey from E
to L along the red axis has length $\sqrt{(20-0)^2 - (0-0)^2} = 20$, which
is the time that passes on Alice's clock.

> To convert the *length* of Bob's (or Alice's) journey to the amount
> of *time* that passes on his clock, you've got to divide the length
> by c (the speed of light). But once again, we're using units where
> $c = 1$, so we can ignore that step.

Note that in spacetime, a straight line (such as Alice's) is always
the *longest* distance between two points!

Students often find it strange and mysterious that Alice and Bob
can part ways at E, meet up again at L, and find that their clocks
have advanced by different amounts.

But if Ted and Carol part ways in Cedar Rapids and meet up again
in Duluth, traveling along very different routes, you probably won't

find it strange or mysterious when I tell you that their odometers have advanced by different amounts.

You'll know you really understand relativity when the difference between Alice and Bob's clock readings seems as natural as the difference between Carol and Ted's odometer readings.

Chapter 18

What We *Can't* All Agree On: The Future and the Past

You might have noticed a problem with the definition of the spacetime interval, which is defined by the formula

$$\sqrt{(\Delta t)^2 - (\Delta x)^2}$$

Namely, If $(\Delta t)^2 < (\Delta x)^2$ (or, equivalently, if $|\Delta t| < |\Delta x|$, or, also equivalently, if $\Delta x / \Delta t > 1$), then the spacetime interval is the square root of a negative number.

For example, if, in Alice's coordinates, E is at $(x = 0, t = 0)$ and F is at $(x = 2, t = 1)$, then the interval from E to F is $\sqrt{1^2 - 2^2} = \sqrt{-3}$.

What should we make of this? If you know something about complex numbers, you might be tempted to say there's no problem at all — the square root of -3 is just $i\sqrt{3}$, where i is a square root of -1. Unfortunately, this doesn't work out so well. Among other difficulties, the complex numbers provide us with *two* square roots of -3 (namely $i\sqrt{3}$ and $-i\sqrt{3}$) and there's no obvious reason to choose one over the other. So, we prefer to say that the *interval* from E to F is undefined, although the *square* of the interval is a perfectly well-defined -3.

When the square of the interval between E and F is positive, we say that the events E and F are *timelike separated*. When the square of the interval between E and F is negative, we say that the events

E and F are *spacelike separated*. When the square of the interval between E and F is zero, we say that E and F are *lightlike separated*. (The reasons why we choose these words will, I hope, become a little clearer shortly. For now, just take them as definitions.)

In the following graph, all events in the shaded region are timelike separated from E. All events in the unshaded region are spacelike separated from E. All events on the gold lines are lightlike separated from E:

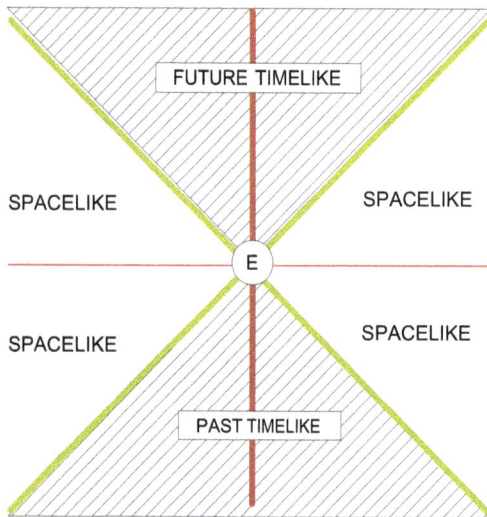

If Alice is present at E, with the indicated boldfaced worldline and horizontal line of simultaneity, then she will say that all events above the horizontal line occur in the future and all events below the horizontal line occur in the past. Therefore, we've divided the timelike region into two parts — the future timelike region and the past timelike region.

Now suppose that Bob is also present at event E but in motion with respect to Alice. We've added his worldline and his line of simultaneity:

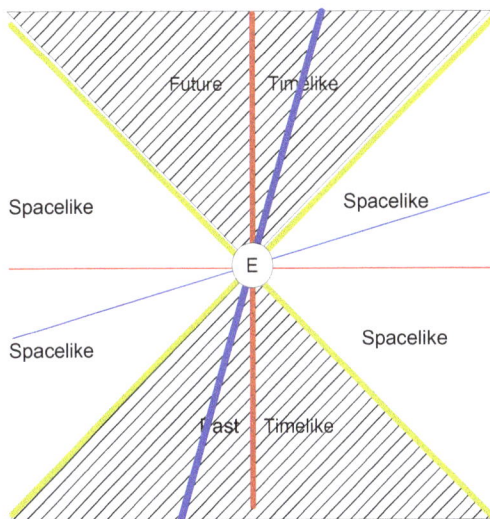

Let's talk about what Bob and Alice do and don't agree on:

- We've already seen that the spacetime interval (or the squared spacetime interval) between any two events is an invariant of the Lorentz transformations. That's a fancy way of saying that Bob and Alice always compute the same squared interval between any two events.

- Because the definitions of *timelike*, *spacelike* and *lightlike* depend only on the squared interval, Bob and Alice must therefore agree about which events are timelike separated from E, which are spacelike separated, and which are lightlike separated. In other words, Bob and Alice completely agree about where the gold lines belong, and which regions should be shaded.

- *Within the shaded timelike region*, Alice and Bob fully agree about which events are in the future and which are in the past. This is because the "future timelike" region lies entirely above Alice's line of simultaneity *and* entirely above Bob's line of simultaneity — that is, they agree that all these events are in the future. Likewise, the "past timelike" region lies entirely below Alice's line of simultaneity *and* entirely below Bob's line of simultaneity — that is, they agree that all these events are in the past.

- However, within the spacelike region, Bob and Alice can disagree about which events are in the future and which are in the past. Consider, for example, the event F, which lies in the future according to Alice, but the past according to Bob; or the event J, which lies in the past according to Alice and the future according to Bob:

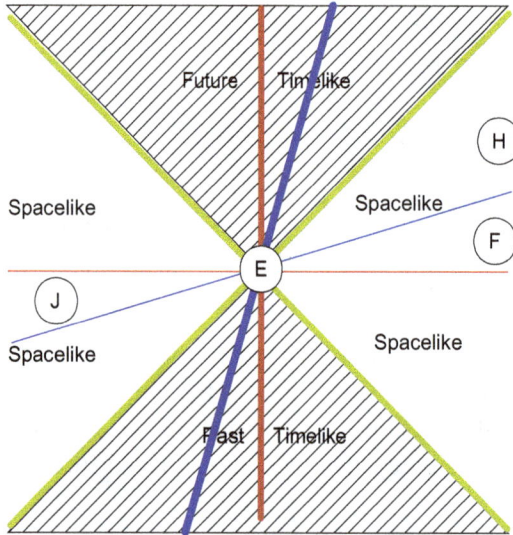

In fact, for *any* event that is spacelike separated from E, we can always come up with an observer who will disagree with Alice about whether that event is in the future or the past. Take, for example, event H in the above graph, which lies in the future according to both Alice and Bob. We can always imagine Jeter, also present at event E, with the green worldline and line of simultaneity, for whom H lies in the past:

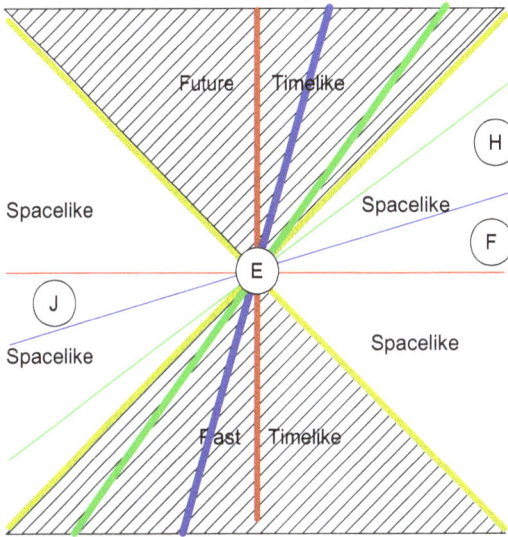

Exercise 18.1. In the graph above, there are 4 labeled events (E, F, H, J). List these events in the order they occur according to Alice (with the red frame), according to Bob (with the blue frame) and according to Jeter (with the green frame).

To summarize:

> **If F is timelike separated from E, then any two observers, both present at E, will agree when asked whether F lies in the past or in the future.**

> **If F is spacelike separated from E, then some pair of observers, both present at E, will disagree when asked whether F lies in the past or in the future.**

Chapter 19

Racing Against Light
(Spoiler: You'll Always Lose)

One Sunday afternoon, Alice was standing on a street corner when Bob went whizzing by. "Hey, Bob!", she called out. "It looks like we both missed Jeter's party yesterday." And Bob yelled back, "Don't be silly, Alice — Jeter's party isn't till tomorrow."

Here is Bob and Alice's meeting (at E) and Jeter's party (at J):

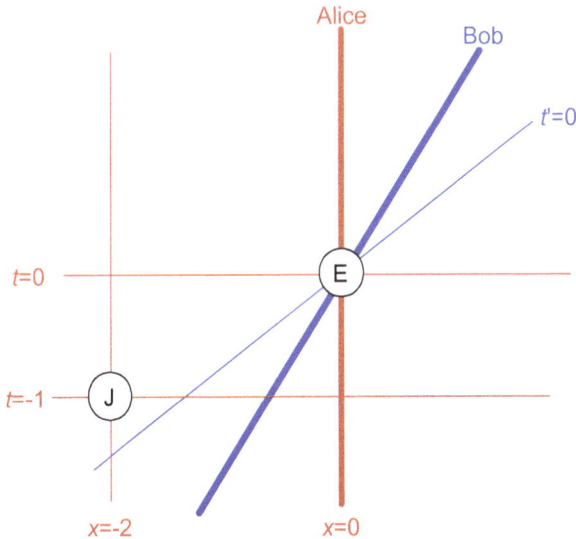

A quick remark: We've seen in the previous chapter that if Bob and Alice (at E) disagree about whether Jeter's party (J) is in the past or the future, then J must be spacelike separated from E. And indeed, you can check that this is so: The squared interval between those events is $(\Delta t)^2 - (\Delta x)^2 = 1 - 4 = -3$, which is negative.

To continue our story: Bob says "Don't be silly, Alice — Jeter's party isn't till tomorrow." And Alice replies "Don't *you* be silly. Jeter already emailed me the party pictures. Do you want to see them?"

If Alice really has those pictures, she can prove Bob is wrong. But according to the First Principle of Relativity, Bob's frame is every bit as legitimate as Alice's, so we know she *can't* prove him wrong. Therefore, she can't have the pictures. In other words, we know she's bluffing.

In fact, Alice can't even know for sure yet that there *is* a party. All she can know (on the basis of some message she received in the past) is that a party was planned. Of course, assuming the party does happen, she might eventually get confirmation by, say, email, but that email can't yet have arrived.

Likewise, Alice cannot have *any information about what happened at Jeter's party*, because according to Bob's perfectly legitimate description, Jeter's party hasn't happened yet.

It follows that no carrier of information — no email, no radio signal, no carrier pigeon and no human being — can travel from Jeter's party at J to Alice's meeting with Bob at E.

What would it take to travel from J to E? It seems that all you'd have to do is hop in your spaceship and fly fast enough to travel a distance 2 in time 1. That is, to get from J to E, all you need to do is fly (according to Alice) at speed 2 — that is, twice the speed of light. But we've just agreed that nobody can get from J to E. Therefore *it must be impossible* for anyone — or for information in any form — to travel (according to Alice) at twice the speed of light.

But of course there's nothing special about Alice, so it must be impossible for anything to travel, according to *anyone*, at twice the speed of light.

A similar argument shows that no information can travel at 12 times the speed of light, or 1.1 times the speed of light, or any other speed greater than lightspeed. In detail:

- Suppose that Bob starts at some event J and travels (according to Alice) at some velocity $v > 1$ until he reaches some event E.
- Let Δt and Δx be the time difference and spatial difference between J and E, according to Alice.
- Then $\Delta x / \Delta t = v > 1$, so $\Delta x > \Delta t$.
- Therefore, the squared interval between J and E is $(\Delta t)^2 - (\Delta x)^2 < 1$.
- Therefore, J and E are spacelike separated.
- Therefore, some observer says that E occurs before J. (Remember from Chapter 18 that there is always some such observer whenever two events are spacelike separated.)
- We know that Bob cannot prove this observer wrong. But traveling from J to E does prove that E comes after J. So, contrary to our original assumption, Bob cannot travel from J to E after all.

Exactly the same argument shows that no information can travel faster than light in the leftward direction, which makes velocities of -2 or -1.3 just as impossible as velocities of 2 or 1.3. **Anything that carries information must travel at a velocity v such that $|v| \leq 1$.**

What about traveling exactly *at* the speed of light? This too turns out to be impossible for any observer. (Though it's obviously not impossible for a light beam.) Here's why: If Bob travels at velocity 1 with respect to Alice, then both his worldline and his lines of simultaneity have slope 1. (Remember that his worldline has slope $1/v$ and his lines of simultaneity have slope v.) In other words, his worldline *is* a line of simultaneity, which means that it looks to Bob like every event in his life takes place at exactly the same moment! This is something that Bob would surely notice, and it would force him to conclude that he's moving (because, after all, this is not how the world looks to an ordinary stationary observer!). But according to the First Principle of Relativity, Bob has every right to consider himself stationary, and nothing can prove he's wrong. So, travel *at* the speed of light is also impossible.

Conclusion: **If Bob moves at velocity v with respect to Alice, then $|v|$ is *strictly* less than 1.** In other words, Bob can't move faster than light, and he can't move exactly as fast as light either.

Chapter 20

Relativity at Everyday Speeds

After all you've learned, you might be tempted to believe that the effects of relativity are always negligible when people travel at everyday speeds. After all, Alice's lines of simultaneity are all horizontal, while Bob's all have slope v. If v is very small (that is, if Bob is not traveling close to the speed of light), then his lines of simultaneity are near horizontal — so his frame and Alice's look nearly identical.

The problem with this argument is that even when a horizontal line and a near-horizontal line cross at the origin, they can get very far apart when you travel far enough away from the origin. So, even if Bob's speed relative to Alice's is in the everyday range, there can still be quite substantial differences in the way they describe far-distant events.

To illustrate, suppose that Alice (heading west) and Bob (heading east) are out driving. They pass each other at a relative speed of 70 mph. (This might mean that each of their speedometers shows 35 mph, or that Alice's shows 50 mph while Bob's shows 20 mph, or that Alice is parked while Bob's speedometer shows 70 mph.) As they pass, Alice waves to Bob. She's smiling, because she's thinking about her beloved cousin Jeter, who lives in a distant galaxy about 400 million light-years west of the Earth and is turning 20 years old just at that moment.

At the moment Alice waves to him, how old does Bob say Jeter is? We don't need any new ideas to figure this out; we can just calculate, using what we already know.

We'll start by looking at things from Alice's frame, in which she is standing still and Bob is traveling eastward at 70 mph. That's just about .0000001 times the speed of light. Thus, Bob's lines of simultaneity have slope .0000001, making them very nearly horizontal. But in order to see what's going on, we'll draw a picture that's very much not to scale and indicate Bob's line of simultaneity in light blue, far enough from horizontal so you can see the difference.

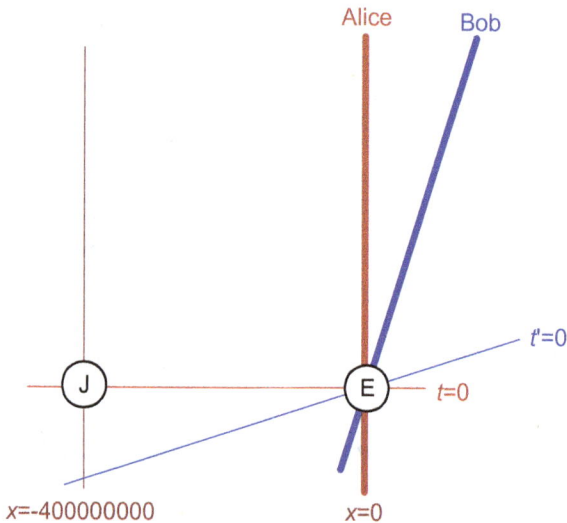

E is the event where Alice waves to Bob; we'll call its coordinates $(x = 0, t = 0)$. J is Jeter's 20th birthday party, which Alice says occurs at $(x = -400000000, t = 0)$. You can see that J is *above* Bob's line of simultaneity, so according to Bob, J lies in the future. To see how *far* he says it lies in the future, we can use the Lorentz transformation

$$t' = \frac{t - vx}{\sqrt{1 - v^2}} = \frac{0 - (.0000001)400000000}{\sqrt{1 - .0000001^2}} \approx 40$$

In other words, Bob says that Jeter's 20th birthday party will take place in 40 years — which means that according to Bob, Jeter *has not even been born yet* and won't be born until 20 years from today.

If we had drawn the picture to scale, Bob's blue line of simultaneity would have been extremely close to horizontal — but the distance from J to E would have been extremely great. For points very close to E, the difference between Bob's coordinates and Alice's is negligible. But the very slight slope of Bob's blue line, extended over the enormous distance from E to J, causes J to lie substantially above the line and therefore causes Bob and Alice to disagree quite substantially — in this case by about 40 years — about its time coordinate.

So, the "negligibility" of relativistic effects requires, at least, that we focus on phenomena involving not just everyday speeds, but also everyday distances.

Here's, the same picture, with one point added: Jeter's birth, at B. If Jeter is stationary with respect to the earth, and if Alice is traveling westward on the earth at 35 mph, then Jeter is traveling eastward at 35 mph with respect to Alice, so his worldline is not quite vertical, but closer to vertical than Bob's. You can see that B is below Alice's (red) line of simultaneity, but above Bob's (blue) line — which is why they disagree about whether it's already happened.

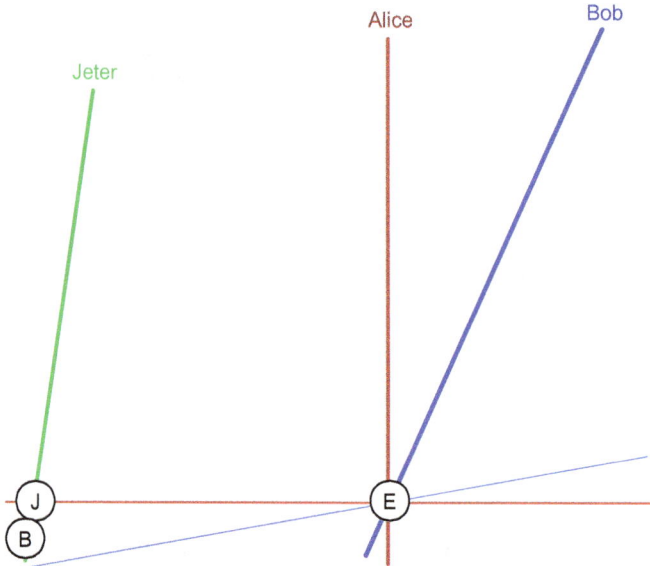

If Bob claims Jeter hasn't been born yet, why can't Alice prove him wrong by showing him one of Jeter's baby pictures? If you've understood the preceding chapter, you already know the answer: Jeter's baby pictures were taken (according to Alice) just 20 years ago — but they were taken 400 million light-years away, which means they'll take at least 400 million years to get here. Alice can't yet *have* any of those baby pictures. Once again, to escape the paradox, we are forced to accept the principle that nothing can travel faster than light.

Chapter 21

Adding Velocities

Alice has just noticed Bob passing by her, heading east at speed v. (Of course, Bob says that it's Alice who's moving at speed v to the west.) Bob has just noticed that Jeter is moving past him, heading east at speed w. At what speed does Alice say Jeter is moving?

To someone who doesn't know anything about relativity, this would appear to be very easy to answer: If Bob's in a train moving past Alice at 50 mph and Jeter is on a scooter moving past Bob at 10 mph in the same direction, then Alice must say that Jeter is moving at 60 mph. More generally, we've assumed that

- Bob passes Alice at speed v
- Jeter passes Bob at speed w

It would seem to follow that Jeter passes Alice at speed $v + w$. But we are about to see that this is only approximately true, though the approximation is excellent at everyday speeds.

Here are Alice's, Bob's and Jeter's worldlines, in boldfaced red, blue and green.

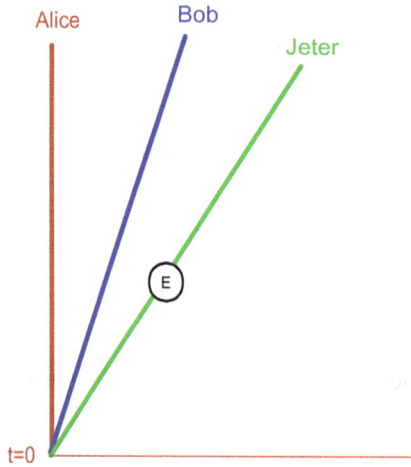

We've made Alice's worldline vertical. But there's nothing special about Alice. We could just as well have made Bob's worldline vertical, and then the diagram would have looked like this:

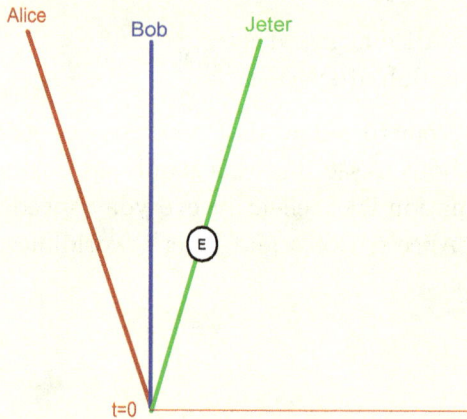

All of the calculations to follow would be unaffected by this choice.

Now we can reason as follows:

- Pick a point E on Jeter's worldline. Alice assigns some coordinates (x, t) to that point. According to Alice, then, Jeter travels distance x in time t, so his speed must be x/t. That's what we need to compute.
- For the same point E, Bob assigns some coordinates (x', t'), so according to Bob, Jeter's speed is x'/t'. But we've already assumed that Bob says Jeter moves at speed w, so we know that $x'/t' = w$, or

$$x' = wt'$$

- We can use the Lorentz transformation to get Alice's coordinates (x, t) in terms of Bob's coordinates (x', t'). Remember, though, that as far as Bob is concerned, Alice is traveling in the negative direction (i.e. westward) so her velocity is $-v$. Then the Lorentz transformation gives

$$x = \frac{x' + vt'}{\sqrt{1 - v^2}} \qquad t = \frac{t' + vx'}{\sqrt{1 - v^2}}$$

so that

$$\frac{x}{t} = \frac{x' + vt'}{t' + vx'} = \frac{wt' + vt'}{t' + vwt'} = \frac{v + w}{1 + vw} \tag{21.1}$$

This, then, is the speed at which Alice says Jeter is traveling. The formula is called the *velocity addition formula*.

Let's apply this formula to the situation from the beginning of this chapter, where Bob is traveling at $v = 50$ mph with respect to Alice, while Jeter is traveling at $w = 10$ mph with respect to Bob. The ordinary non-relativistic approximation is that Jeter is traveling at 60 mph with respect to Alice. To see how good this approximation is, consider that in units where the speed of light is equal to 1, 50 mph is .0000000000085 and 10 mph is .0000000000017. According to formula 21.1, we first add v and w (getting 60 mph) and then divide by $1 + vw$, which is about 1.00000000000000000000000015.

Thus, Jeter passes Alice at a speed (in miles per hour) of

$$\frac{60}{1.0000000000000000000000015} = 59.9999999999999999999132$$

So, the usual approximation of 60 mph is pretty good.

Exercise 21.1. Jeter is standing on earth, watching Alice and Bob fly past him in opposite directions, both at speed 1/2. They both pass him just as all three of their clocks strike time 0, with Alice headed west and Bob headed east. Jeter says that there are mileposts one light-year from him in each direction.

(a) According to Jeter, at what time does Bob reach the eastern milepost?

(b) According to Bob, at what time does Bob reach the eastern milepost?

(c) According to Alice, at what time does Bob reach the eastern milepost? [Let E be the event where Bob reaches the milepost. Write down the coordinates of E in Jeter's frame, then transform those coordinates to Alice's frame.]

(d) Use your answers to parts (b) and (c) to determine the speed of Bob's clock according to Alice.

(e) At what speed is Bob traveling according to Alice? [Use the velocity addition formula.]

(f) Use your answers to part (e) and Exercise 13.2 to determine the speed of Bob's clock according to Alice.

   If your answers to parts (d) and (f) don't agree, then you have done something wrong!

Now continue to suppose Bob travels at velocity v with respect to Alice and Jeter travels at velocity w with respect to Bob, so that Jeter travels at velocity $\frac{v+w}{1+vw}$ with respect to Alice. Then with a little algebra, you can check that this is true:

$$\text{If } |v| \leq 1 \quad \text{and} \quad |w| \leq 1, \quad \text{then} \left| \frac{v+w}{1+vw} \right| \leq 1.$$

This is reassuring, because we know from preceding chapters that *anyone's* speed with respect to *anyone* must be ≤ 1. So, if it were possible for $\left| \frac{v+w}{1+vw} \right|$ to be greater than 1, there would be something dreadfully wrong with our theory.

PART VI

UNDERSTANDING SPACE: DISTANCES AND LENGTHS

Chapter 22

Distances

Once again, Alice is standing on a street corner just as Bob flies by her, heading eastward. He's traveling at $v = 3/5$ (that is, at $3/5$ lightspeed), and they both agree to assign coordinates ($x = 0$, $t = 0$) to the event "Bob passes Alice".

Bob is heading toward Barnard's Star, which is, as you know, at rest with respect to earth, and, by Alice's calculation, 6 light-years away. How far does Bob say it is to Barnard's Star?

Here is Alice's worldline in red, Barnard's Star's worldline in black, and Bob's in blue:

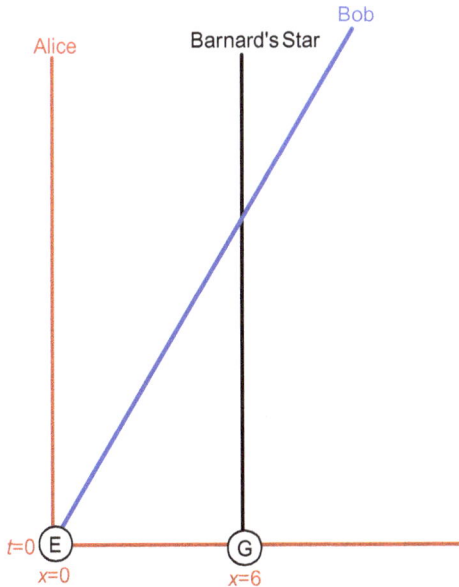

As Bob passes Alice at E, she says that "Barnard Star, right now" is the event G, with coordinates $(x = 6, t = 0)$. Using the Lorentz transformation (we've done this calculation before!) we find that Bob says event G takes place at location

$$x' = \frac{x - vt}{\sqrt{1 - v^2}} = \frac{6 - (3/5) \cdot 0}{\sqrt{1 - (3/5)^2}} = 7.5$$

That is, Bob says that event G is 7.5 light-years away.

But 7.5 light-years is *not*, according to Bob, the current distance to Barnard's Star. It's the distance to *event* G, which, according to Bob, took place at Barnard's Star sometime in the past. (In fact, at time $t' = (t - xv)/\sqrt{1 - v^2} = -4.5$).

To compute the *current* distance to Barnard's Star according to Bob, we have to add Bob's line of simultaneity, and locate the event H, which, according to Bob at event E, is happening on Barnard's Star *right now*:

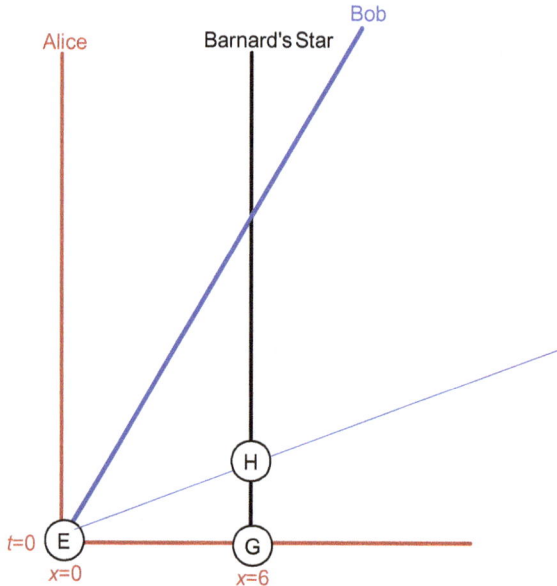

We want to know how far it is to Barnard's Star in Bob's frame, which means we need to know the coordinates Bob assigns to H. We'll solve this problem by *first* figuring out the coordinates that

Alice assigns to H, and then using the Lorentz transformations to convert to Bob's coordinates.

So, what coordinates does Alice assign to H? Answer: The blue line of simultaneity has slope equal to Bob's velocity $3/5$, and it goes through the origin, so its equation is $t = (3/5)x$. Since $x = 6$, we have $t = (3/5) \cdot 6 = 3.6$.

Now we can apply the Lorentz transformation to get

$$x' = \frac{x - vt}{\sqrt{1 - v^2}} = \frac{6 - (3/5) \cdot 3.6}{\sqrt{1 - (3/5)^2}} = 4.8$$

So, when Bob passes Alice, he says "Right now, I am at event E and event H is taking place on Barnard's Star. Event H is 4.8 light-years away, so Barnard's Star is currently 4.8 light-years away."

The following picture shows relevant gridlines from Alice and Bob's frames, incorporating the calculations we've just completed:

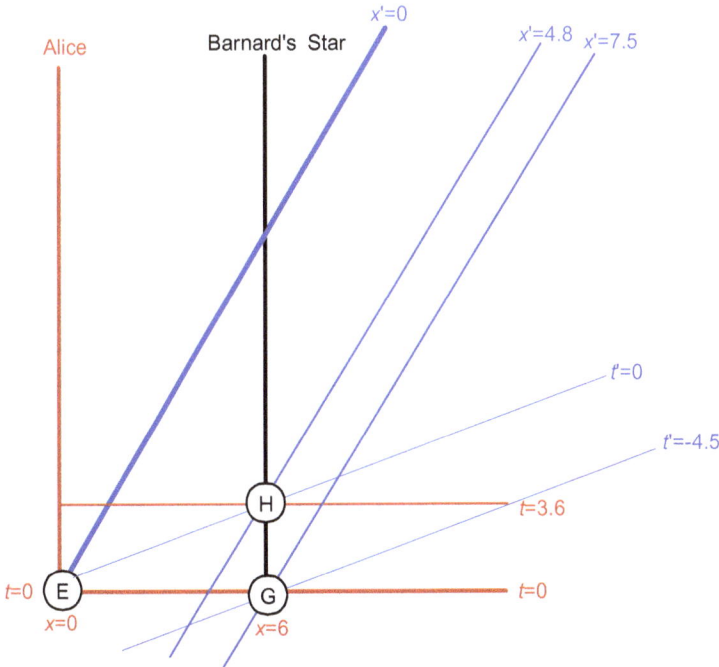

Note that in order to calculate the distance to Barnard's Star in Bob's frame, we had to make *two* corrections to Alice's calculation.

First, we had to replace the event G (which Alice says is "right now") with the event H (which Bob says is "right now"). Second, after figuring out the coordinates that Alice assigns to H, we had to use the Lorentz transformation to switch from Alice's coordinates to Bob's.

If G and H both take place on Barnard's Star, how can Bob say one is farther than the other? Answer: In Bob's frame, Barnard's Star is moving toward him, and has always been. Event G took place 4.5 years ago, when Barnard's Star was quite a bit farther away.

In fact, let's do a quick reality check: Barnard's Star is moving towards Bob at speed 3/5. Event G took place 4.5 years ago, so the star must have moved (3/5)(4.5)=2.7 light-years closer in the meantime. This accounts for the 2.7 light-year difference between the distance to event H and the distance to event G.

Exercise 22.1. Suppose that Bob passes Alice at velocity v en route to a distant galaxy that is at rest with respect to Alice. She says that the distant galaxy is D light-years away. How far away does Bob say the distant galaxy is? (Answer: $D\sqrt{1-v^2}$. That is, according to Alice, Bob underestimates the distance by a factor of $\sqrt{1-v^2}$.)

Exercise 22.2. The Andromeda galaxy is 2.5 million light-years from earth (and at rest relative to earth). Bob travels from earth to Andromeda at a very high speed. As soon as he takes off in his rocket ship, he calculates that Andromeda is only 20 years from earth, so he is able to complete his journey in just a little over 20 years.

His friend Alice is confused: "Wait a minute, Bob. Just before you took off, you agreed that Andromeda was 2.5 million light-years away. Just afterward, you said it's only 20 light-years away. Did Andromeda really move 2.5 million light-years in the blink of an eye?"

Help to relieve Alice's confusion by drawing a spacetime diagram that shows Bob's (2.5 million light-year long) ruler at time 0 according to Alice, the same ruler at time 0 according to Bob (immediately after he's taken off) and the worldline of the "20 light-year" mark on that ruler.

Chapter 23

Lengths

Until now, we've talked about *the* location of Bob's ship at any given moment (either according to Bob or according to Alice). But a real-world spaceship does not occupy just one location. The front of the ship and the back of the ship are never in exactly the same place at the same time.

So let's account for that. Bob is once again flying his spaceship eastward past Alice at velocity $v = 3/5$. His pilot seat is located in the back of the ship — at the west end. His hood ornament (a dragon) is attached to the front. He (and the back end of the ship) pass Alice at event E, which they both label $(x = 0, t = 0)$. Here are Alice's worldline in red, Bob's in blue, and the dragon's in green:

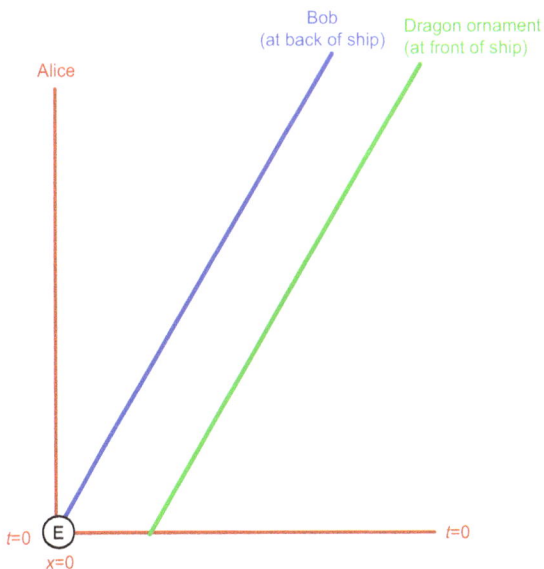

As Bob passes Alice at point E, let's ask them each this question: *How long is Bob's ship?*

The answer, surely, is that the length of the ship is equal to the distance from *the back of the ship right now* to *the front of the ship right now*. But those words mean one thing to Alice and another to Bob.

According to Alice (at time 0 by her watch) *the back of the ship right now* is at E and *the front of the ship right now* is at F in the picture below. We'll assume that F happens to be at $x = 1$, so Alice reports that the length of the ship is 1:

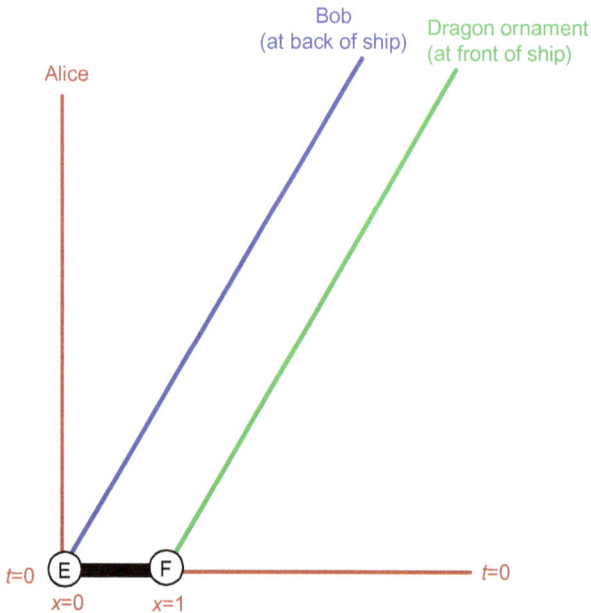

(The black line segment, according to Alice, represents the entire ship at time 0.)

According to Bob (at time 0 by his watch) *the back of the ship right now* is at E and *the front of the ship right now* is at G in the picture as follows:

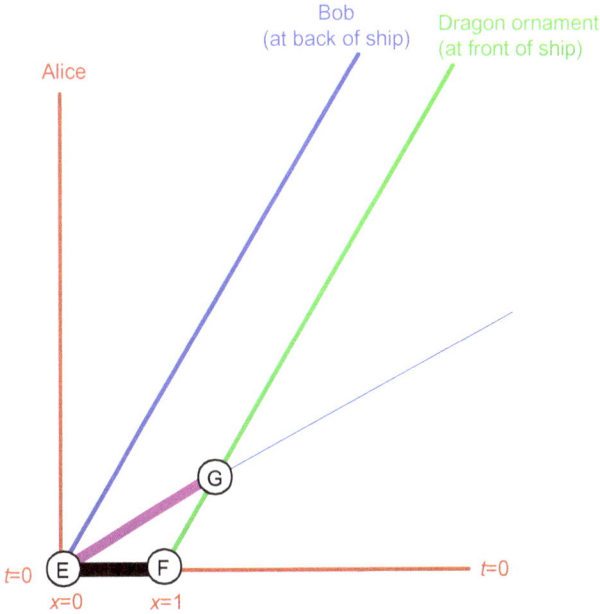

We need the coordinates of point G. For this, note that

- G is on Bob's line of simultaneity through the origin, which has slope $3/5$ and therefore has the equation $t = (3/5)x$.
- G is also on the dragon ornament's worldline, which passes through $(x = 1, t = 0)$ and has slope $5/3$, so that its equation is given by $t = (5/3)(x - 1)$.
- Therefore, we can solve for the coordinates of G (in Alice's frame) by solving the equations

$$t = (3/5)x = (5/3)(x - 1)$$

giving

$$x = 25/16 \quad t = 15/16$$

- To find the coordinates of G in Bob's frame, we use the Lorentz transformations

$$x' = \frac{x - vt}{\sqrt{1 - v^2}} = \frac{\frac{25}{16} - \left(\frac{3}{5} \times \frac{15}{16}\right)}{\sqrt{1 - (3/5)^2}} = 5/4$$

$$t' = \frac{t - vx}{\sqrt{1 - v^2}} = \frac{\frac{15}{16} - \left(\frac{3}{5} \times \frac{25}{16}\right)}{\sqrt{1 - (3/5)^2}} = 0$$

The last equation is a good reality check: Bob says G and E are simultaneous, so he had better say that G takes place at time $t' = 0$. The preceding equation tells us that Bob says event G — *the front of the ship right now, at time $t' = 0$*, takes place at location 5/4, so his ship has length 5/4.

In short: If Bob is traveling at velocity v with respect to Alice, and if Alice says that Bob's spaceship has length 1 (in yards, meters, hundreds of yards, or whatever other unit they're using), then Bob must say his spaceship is 25% longer.

The *proper length* of the spaceship is defined to be the length as measured by a traveler on the ship — in this case that traveler is Bob, so the ship has proper length 5/4. Any other observer — such as Alice — will say that the length of the ship is something shorter than its proper length.

This phenomenon — the fact that everything (including Bob's ship) is longer in its own frame than in Alice's (or in any other) — is usually called the *Lorentz contraction*.

The phrase "Lorentz contraction" sometimes misleads students into thinking that Bob's spaceship was once longer than it is, and has somehow grown shorter. Nothing of the sort needs to have happened. If Bob is always in motion with respect to Alice, then his ship always has one length in his own frame and a different, shorter, length in Alice's frame. We still speak of a "Lorentz contraction" even if neither length has changed. If Bob's ship changes frames (that is, if it accelerates), then its proper length (that is, its length in its own frame) might or might not change depending on the details of exactly how Bob accelerates (we'll see this in the next two chapters). What remains true is that after any changes, the ship's new length in Bob's frame must be longer than its new length in any other.

In the preceding chapter, Alice and Bob were measuring the distance between two objects (namely earth and Barnard's Star), both of which are at rest with respect to Alice. We saw that Bob's measure of the distance is $\sqrt{1-v^2}$ times Alice's measure. In this chapter, Alice and Bob were measuring the distance between two objects (namely the front and back of Bob's ship), both of which are at rest with respect to Bob. In other words, the roles of Bob and Alice are reversed from the preceding chapter, which tells us that Alice's measure of the ship length must be $\sqrt{1-v^2}$ times Bob's measure. This reasoning yields all the conclusions of the present chapter and makes the chapter itself unnecessary! But it's always good to do things more than one way, to make sure the answers agree.

Everything we've said about Bob's spaceship applies equally well to everything on board the ship. Everything Bob is carrying appears shorter to Alice than it does to Bob.

⚡ Our pictures are all two-dimensional, with one dimension of time and one of space. A more realistic picture would include two dimensions of space and an even more realistic one would include three dimensions. In such a picture, we'd find that Bob's spaceship, and everything aboard that ship, appears shorter to Alice than to Bob *in the direction of the ship's motion*, but not in other directions. For example, if Bob is holding a meter stick pointed straight ahead in his direction of motion, Alice will say that meter stick is less than a meter long. If he holds the meter stick in a direction perpendicular to his direction of motion, then Alice and Bob will agree on the *length* of the meter stick, but will disagree about its *width*.

Exercise 23.1. Bob, sitting in the back of his spaceship, passes Alice at some velocity v. According to Alice, the length of his ship is L. (In the calculation above, we assumed $v = 3/5$ and $L = 1$; now we are dropping those assumptions.) Find the proper length of his ship. (Answer: $L/\sqrt{1 - v^2}$.)

Exercise 23.2. Repeat the previous exercise, assuming that Bob is sitting in the front of his spaceship instead of the back.

Exercise 23.3. Alice owns a tunnel with proper length 1. Bob owns a spaceship with proper length 1, which is about to enter the tunnel, traveling at velocity 3/5.

(a) According to Alice, how long is the spaceship? (Answer: 4/5.)

(b) According to Bob, how long is the tunnel? (Answer: 4/5.)

(c) Alice has decided to prove that Bob's spaceship is shorter than her tunnel by briefly closing the front and back doors to the tunnel while Bob's spaceship is fully inside — then opening the doors to let him out. "Bob", she says, "you can't deny that your ship was fully inside the tunnel while the doors were closed. If it had been sticking out in either direction, the doors would have hit it." Draw a spacetime diagram showing worldlines for the front and back of the tunnel and the front and back of the ship. Use your diagram to illustrate Alice's argument. Use the same diagram to illustrate Bob's rebuttal. Does Bob have to admit that Alice is right?

Chapter 24

Contraction in Action: Part I

Bob has just bought a shiny new spaceship and is showing it off to Alice. Examining it in the parking lot, they both agree that it's exactly 1 mile long. Now Bob climbs into the ship and takes off, instantly accelerating to $v = 3/5$. At this point, we know from the preceding chapter that Alice and Bob must disagree about the length of Bob's spaceship. Indeed, if Alice says the ship has length L, then Bob must say the ship has length $L/\sqrt{1 - v^2} = (5/4)L$. So, they can't continue to agree about the length of the ship. Whose mind has changed — Bob's or Alice's?

Based on the information given, it's quite impossible to answer this question! That's because we don't know the details of the ship's acceleration. Did the front of the ship jump up to speed first, and then the back a moment later? Or vice versa? Or did every part of the ship jump up to speed all at once?

You might be tempted to say that the natural interpretation of the question is that every part of the ship jumps to speed at once. But then we have to ask *at once according to whom*? If all parts of the ship accelerate at once according to Alice, then they can't have accelerated all at once according to Bob. If they accelerated all at once according to Bob, they can't have accelerated all at once according to Alice.

The figure below adds the assumption that *the entire ship accelerates simultaneously according to Alice*. The worldlines of the front and back of the spaceship are shown in solid red and blue. From time $t = -1$ to $t = 0$, the ship is at rest with respect to Alice, and those worldlines are vertical. At $t = 0$ (by Alice's watch) all parts of the ship take off, so the worldlines of both the front and the back begin sloping to the right, while earthbound Alice continues with the dashed red worldline. The black lines are, according to Alice, the spaceship at times 0, 1, 2, and so forth. Obviously, these all have length 1 mile in Alice's coordinates. Therefore, Alice says that the spaceship continues to have length 1 after Bob takes off.

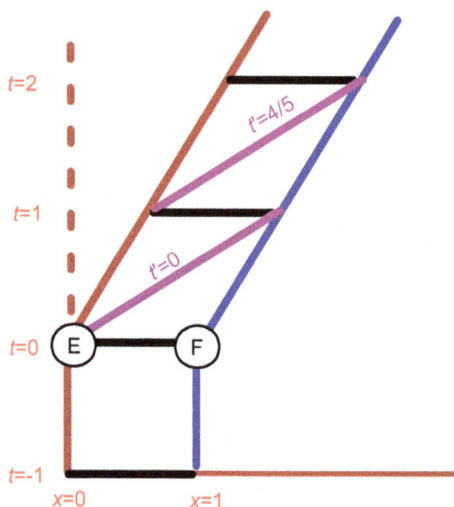

The purple lines are, according to Bob, the spaceship at times $t' = 0$, $t' = 4/5$, and so on. As we calculated in the previous chapter, Bob says that these lines all have length 5/4. In other words, his spaceship is longer now.

How does Bob explain this expansion? Note that once Bob is in the air, he says that event F ("front of ship takes off") occurred *before* event E ("back of ship takes off"). (In fact, you can calculate that he says F takes place at $t' = -3/4$, while event E takes place at time $t' = 0$.) There was, therefore, a period of time in which the front of the ship was moving forward, while the back was standing still. So — according to Bob — it's no surprise that his ship has stretched.

In short:

- Alice says: "Bob bought this new spaceship that's one mile long. At time 0, the whole ship started moving, all at once, and it's moving still. At each moment, the front and back of the ship were moving at identical speeds — so the distance from the front to the back never changed. Of course, the spaceship is still one mile long."

- Bob says: Once upon a time, my spaceship was one mile long. Then, at time $t' = -3/4$, the front of the ship started moving, and gradually other parts started moving too, until finally at time $t' = 0$, the back started moving. Of course that caused the ship to stretch, so now it's 25% longer."

In short:

- In Alice's frame, Bob's ship retains its length when it takes off.
- In Bob's frame, his ship expands when it takes off.
- Therefore Bob's ship is longer in its own frame than in Alice's (or in any other). In other words, Bob's ship remains Lorentz-contracted.

You might find it puzzling that after Bob's ship has *expanded*, we say that the ship is Lorentz *contracted*. That's because the Lorentz contraction refers *not* to a comparison between Bob's ship at one time and Bob's ship at another time; instead it refers to a comparison between Bob's ship in Alice's frame and Bob's ship in Bob's frame — and it expresses this comparison from Alice's point of view. In Bob's post-takeoff frame, his ship expands, which makes it *shorter* in Alice's frame than in Bob's.

An Application: As you know, Bob and Alice disagree about the length of Bob's ship only because they are using different frames to describe exactly the same events. So you might think that in order to understand what's going on, it doesn't matter which frame we use. Yet it turns out that some very important events are quite easy to understand in one frame and quite difficult to understand in the other.

A case in point: From Bob's point of view, his ship has stretched. Because he's moving at speed $3/5$, his ship has stretched by a factor of $1/\sqrt{1 - (3/5)^2} = 5/4$. If he'd been traveling faster, it would have stretched even more. But there's a limit to how much a spaceship can stretch before it snaps into pieces. Therefore:

> **If Bob accelerates to a high enough speed, and does so in such a way that, according to Alice, the acceleration occurs simultaneously at every part of the ship, then Bob's ship must fall apart.**

Now of course when Bob's ship falls apart, Alice can hardly fail to notice. But if she insists on explaining *why* it fell apart using only her own frame, she'll have some hard work to do, analyzing the strains imposed on the ship's components by the forces that caused it to accelerate. If she's open-minded enough to see things from Bob's point of view, she need only note that things tend to break when you stretch them too far.

Chapter 25

Contraction in Action: Part II

In the preceding chapter, every part of Bob's ship accelerates *simultaneously according to Alice*. Now let's suppose instead that every part of the ship accelerates *simultaneously according to Bob*.

More precisely: Bob starts off sharing Alice's frame, and then changes frames when he accelerates. So, our assumption is that every part of the ship accelerates simultaneously according to Flying Bob, as opposed to Resting Bob, who agrees with Alice.

Here's the picture:

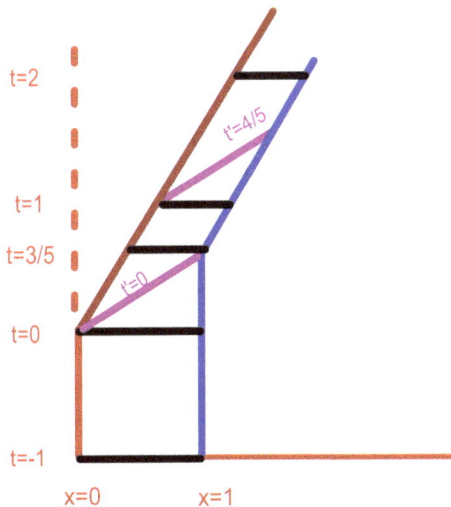

Each of the black line segments is the spaceship at a given instant according to Alice. (That is, these segments connect the front of the ship to the back of the ship, and run along Alice's lines of

133

simultaneity.) Each of the purple line segments is the spaceship at a given instant according to Flying Bob.

Because Flying Bob says that the ship's stern took off at time $t' = 0$, the red worldline of the ship's stern must have a kink at $t' = 0$. Because he says that the ship's prow took off at the same time $t' = 0$, the blue worldline of the ship's prow must have a kink at $t' = 0$. As you can see from the picture, this forces the black lines after $t = 0$ to be shorter than the black lines at $t = 0$ and earlier. That is, Alice says that after $t = 0$, the ship has shrunk.

In short:

- Bob says: My ship is a mile long. At time $t' = 0$, the whole ship took off at once. It's length, of course, didn't change. Why would it?
- Alice says: Bob's ship was once a mile long. At time $t = 0$, the back of the ship started moving forward. Later on, at $t = 3/5$, the front started moving forward. Because the back started moving before the front did, the ship got squashed. Now it's only 4/5 of a mile long.

Once again, as in the preceding chapter, Bob's ship is longer in its own frame than it is in Alice's, which is exactly the phenomenon we called *Lorentz contraction*.

Exercise 25.1.   Alice and Bob are riding in the front and back of a train 100 light-seconds long, traveling at 7/10 the speed of light. They simultaneously drop rocks out the train's window. According to a ground-based observer, how far apart are the rocks when they are dropped?

Appendix I

Hints and Solutions for All the Exercises

4.1. Bob's coordinates are $x' = (3/5)(2) + (4/5)(1) = 2$ and $y' = (-4/5)(2) + (3/5)(1) = -1$.

4.2. Alice's coordinates are $x = (3/5)(-1) - (4/5)(3) = -3$ and $y = (4/5)(-1) + (3/5)(3) = 1$.

5.1. Alice computes the distance from the dandelion to the tree as $\sqrt{6^2 + 2^2} = \sqrt{40} \approx 6.3$. Bob computes the same distance as $\sqrt{2^2 + 6^2} = \sqrt{40} \approx 6.3$.

Alice computes the distance from the tree to herself as $\sqrt{4^2 + 3^2} = 5$. Bob computes the same distance as $\sqrt{0^2 + 5^2} = 5$.

5.2.

$$\sqrt{(\Delta x')^2 + (\Delta y')^2}$$
$$= \sqrt{(\cos(\theta)\Delta x + \sin(\theta)\Delta y)^2 + (-\sin(\theta)\Delta x + \cos(\theta)\Delta y)^2}$$
$$= \sqrt{(\Delta x)^2 + (\Delta y)^2}$$

9.1. (a) We've computed that $x' = 0$ and $t' = t\sqrt{1 - v^2}$. Inserting these into the Lorentz equations (and remembering to use $-v$ instead of v for the velocity), we get

$$x = \frac{x' + vt'}{\sqrt{1 - v^2}} = vt \quad t = \frac{t' + vx'}{\sqrt{1 - v^2}} = t$$

Both equations are confirming what we already knew, which is further evidence that the Lorentz transformations work as they ought to.

(b) Because the event is on Alice's worldline, and because we know Bob's equation for that worldline, we know that $x' = -vt'$. Now, applying the Lorentz equations, we get

$$x = \frac{x' + vt'}{\sqrt{1 - v^2}} = 0 \quad t = \frac{t' + vx'}{\sqrt{1 - v^2}} = t'\sqrt{1 - v^2}$$

9.2. Red is Alice; blue is Bob:

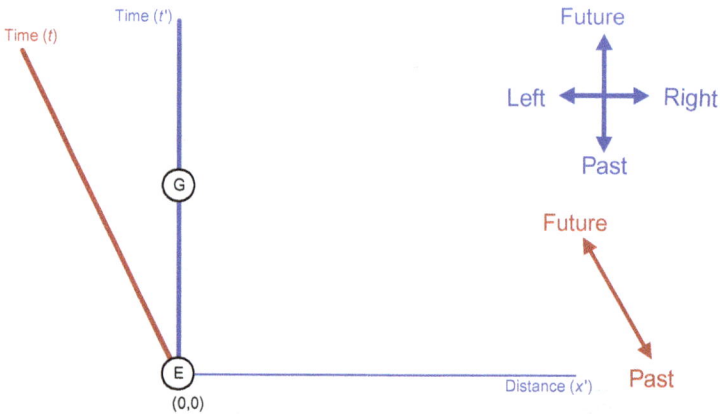

10.1. Use the Lorentz transformation

$$x' = \frac{x - vt}{\sqrt{1 - v^2}} = \frac{2 - (v)(1)}{\sqrt{1 - v^2}}$$

If $v = 1/2$, this becomes

$$x' = \frac{2 - (1/2)(1)}{\sqrt{1 - (1/2)^2}} = \sqrt{3} \approx 1.73$$

13.1. We know that when Bob leaves earth (event E) he says he is 20 years old, and when he arrives at Barnard's Star (event F), he says is 28 years old. We also know (by projecting on the vertical axis) that event J is 64% of the way from E to F. So, at event J, Bob says he is 25.12 years old.

Alternatively, we could have applied the Lorentz transformation. We know that in Alice's coordinates, J is given by $t = 6.4$ and $x = 3.84$. Putting this into the Lorentz transformation gives $t' = 5.12$ (so Bob's age is 25.12).

Bob's line of simultaneity through J (with slope 3/5) crosses Alice's worldline at $t = 4.096$. So when Bob is at J, he says that Alice says that he is 4.096 years into his journey, making her 24.096 years old.

13.2. Here is the picture: Bob is traveling at speed v.

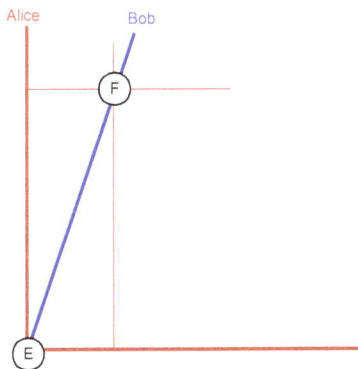

At any given time t, Alice says that Bob has traveled a distance $x = tv$. Therefore, any event on Bob's worldline, such as F, has coordinates of the form (tv, t) for some t. Transforming to Bob's coordinates via the Lorentz transformations, we get

$$t' = \frac{t - xv}{\sqrt{1 - v^2}} = \frac{t - tv^2}{\sqrt{1 - v^2}} = t\frac{1 - v^2}{\sqrt{1 - v^2}} = t\sqrt{1 - v^2}$$

Thus, Alice says that at time t, Bob's clock reads $t\sqrt{1 - v^2}$, and hence his clock is running slow by the factor $\sqrt{1 - v^2}$.

15.1. The key points in the journey are E, F and L in the graph below (identical to the graph on page 69):

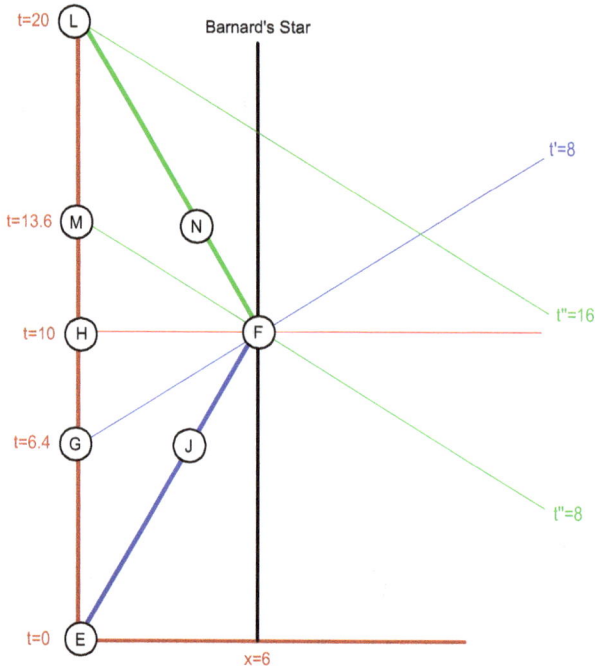

Here are the coordinates of those points in Alice's frame and in Outbound Bob's frame (computed via the Lorentz transformation with $v = 3/5$):

	Alice's Coordinates	Outbound Bob's coordinates
E	$(x = 0, t = 0)$	$(x' = 0, t' = 0)$
F	$(x = 6, t = 10)$	$(x' = 0, t' = 8)$
L	$(x = 0, t = 20)$	$(x' = -15, t' = 25)$

So Outbound Bob says this:

- Bob left on his 20th birthday, when he and Alice set their watches to zero.

- After 8 years, Bob arrived at Barnard's Star. During that 8 years, Bob had aged 8 years, so his clock was running normally. During that 8 years, Alice had aged only 6.4 years, so her clock was running at $6.4/8 = 4/5$ speed.
- After another 17 years (from time $t' = 8$ to time $t' = 25$) Bob returned to earth. During that 17 years, Bob aged an additional 8 years, so his clock was running at $8/17$ speed. During that 17 years, Alice aged an additional $20 - 6.4 = 13.6$ years, so her clock was running at $4/5$ speed.

15.2. Referring to the graph from the previous problem, here are the deltas between relevant events in Alice's frame and in Inbound Bob's frame (computed via the Lorentz transformation with $v = -3/5$, and indicated with double-primes):

	Alice's deltas	Inbound Bob's deltas
E to F	$(\Delta x = 6, \Delta t = 10)$	$(\Delta x'' = 15, \Delta t'' = 17)$
F to L	$(\Delta x = -6, \Delta t = 10)$	$(\Delta x'' = 0, \Delta t'' = 8)$

(We need to use deltas instead of the coordinates of individual points because Alice and Inbound Bob don't share an origin.)

So Inbound Bob says this:

- Bob left Earth on his 20th birthday, when he and Alice set their watches to zero.
- After 17 years, Bob arrived at Barnard's star. During that 17 years, Bob had aged 8 years, so his clock was running at $8/17$ speed. During that 17 years, Alice aged 13.6 years, so her clock was running at $13.6/17 = 4/5$ speed.
- After another 8 years, Bob arrived back at earth. During that 8 years, Bob aged 8 years, so his clock was running normally. During that 8 years, Alice aged an additional $20 - 13.6 = 6.4$ years, so her clock was running at $6.4/8 = 4/5$ speed.

15.3. Here's the picture:

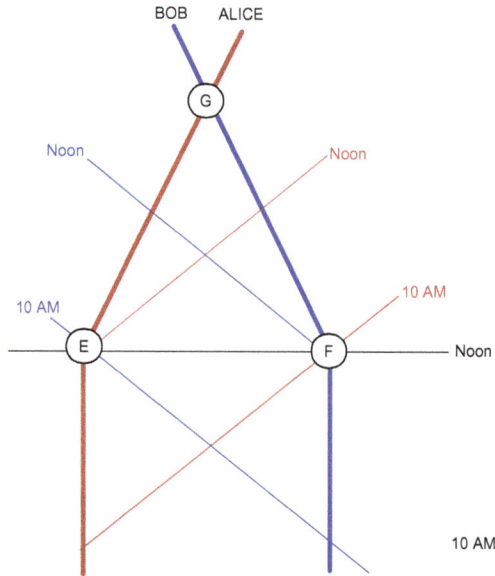

The boldfaced lines are Alice's and Bob's worldlines. The black horizontal line connecting E with F is a line of simultaneity for Alice and Bob as long as their worldlines are vertical (i.e. until they take off). So, the instant before taking off, they can agree that Alice's takeoff (E) and Bob's takeoff (F) occur simultaneously at noon.

But the instant Alice takes off, her frame changes and she has the red lines of simultaneity. Now she says that her own takeoff E took place at noon, but that Bob's takeoff F took place at some earlier time, say 10 AM. Bob says the opposite: His own takeoff took place at noon, and Alice's at 10 AM.

Alice says:

- When I took off at noon, my clock said noon. When Bob took off at 10 AM, his clock said noon. So, his clock started out two hours fast. When we met at G our clocks agreed, even though his started out two hours ahead of mine, so his must have been running slow.

Bob, of course, says the same thing in reverse.

15.4. The graph below shows Alice's (red) outbound and (blue) inbound worldlines on the left and Bob's (blue) outbound and (red) inbound worldlines on the right. The black coordinates are taken from their Mom's description of their journeys.

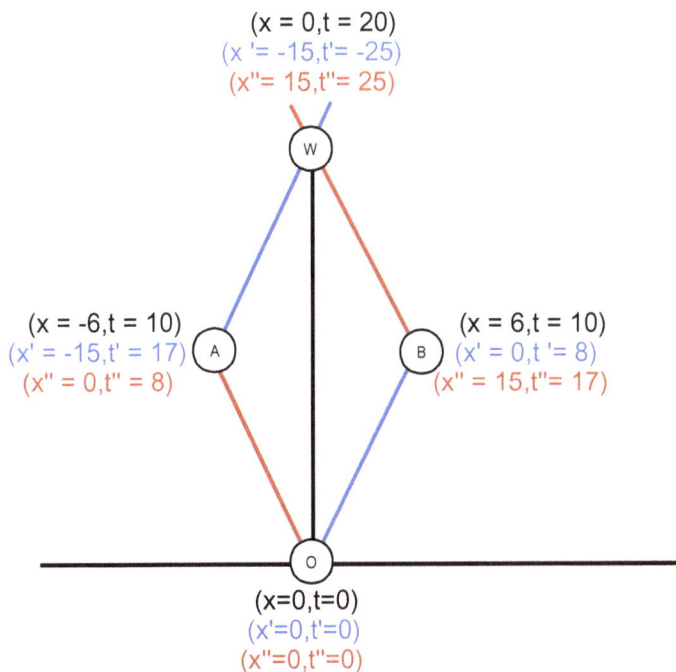

(x = 0,t = 20)
(x '= -15,t'= -25)
(x"= 15,t"= 25)

W

(x = -6,t = 10)   (x = 6,t = 10)
(x' = -15,t' = 17) A   B (x' = 0,t '= 8)
(x" = 0,t" = 8)   (x" = 15,t"= 17)

O

(x=0,t=0)
(x'=0,t'=0)
(x"=0,t"=0)

The blue coordinates at each point are derived from the black coordinates assuming a velocity of $3/5$ and the red coordinates are derived from the black coordinates assuming a velocity of $-3/5$.

When Alice reports her descriptions, she is passing through W on her blue worldline at velocity $3/5$, so she describes the world in terms of the blue coordinates. Using those coordinates, she reports as follows:

(a) "I left earth $\boxed{25}$ years ago, and traveled for $\boxed{17}$ years at speed $\boxed{15/17}$, during which my clock moved at $\boxed{8/17}$ times normal speed, so that it advanced a total of $\boxed{8}$ years. Then I turned around. Since then it's been $\boxed{8}$ years (while I sat still and the earth moved toward me), during which my clock moved at $\boxed{1}$ times normal speed, so that it advanced a total of another $\boxed{8}$ years. That's why my clock now says $\boxed{16}$."

(b) "Bob left earth $\boxed{25}$ years ago, and traveled for $\boxed{8}$ years at speed $\boxed{0}$ (that is, he stood still while the earth moved away from him), during which his clock moved at $\boxed{1}$ times normal speed, so that it advanced a total of $\boxed{8}$ years. Then he turned around and traveled another $\boxed{17}$ years, during which his clock moved at $\boxed{8/17}$ times normal speed, so that it advanced a total of another $\boxed{8}$ years. That's why his clock now says $\boxed{16}$."

16.1. If we take Bob's departure as the origin, then the event "Alice sends signal" takes place at $x = -8$, $t = 0$. As soon as Bob starts traveling toward earth at velocity $v = -4/5$ (the minus sign is because he's traveling *toward* earth), he changes coordinates to $x' = (x - tv)/\sqrt{1 - v^2} = -13.33$ and $t' = (t - xv)/\sqrt{1 - v^2} = -10.67$.

Because Bob always says that his ship is at $x' = 0$, the signal (according to Bob) travels 13.33 light-years to reach him. It travels at the speed of light, and so must take 13.33 years to arrive. It left at time -10.67, so it must arrive at time $13.33 - 10.67 = 2.67$.

(There are many other ways to reach the same conclusion.)

16.2.

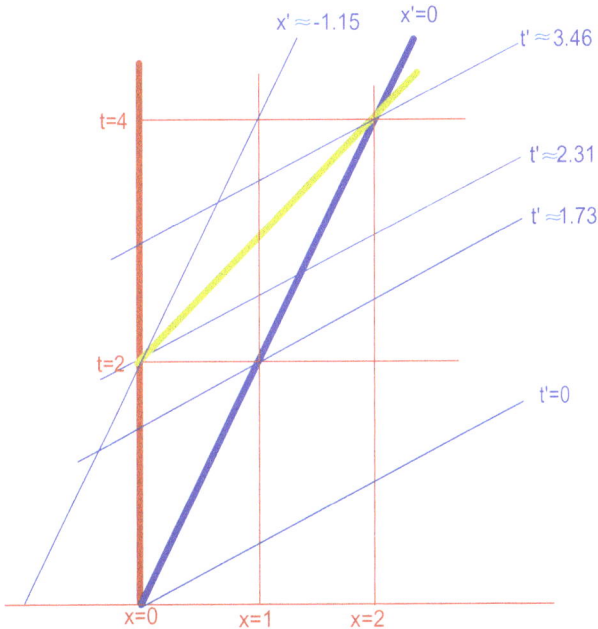

- Alice says: At time 2, when Bob reached location 1 (and his slow clock was reading 1.73), I sent my message. At time 4 (when Bob was at location 2 and his slow clock was reading 3.46), the message caught up with him.
- Bob says: At time 2.31, when Alice (who along with the rest of the earth had been traveling leftward) reached location −1.15 (and when her slow clock was reading 2), she sent me a message. Since I'm not moving, the message (traveling at speed 1) took 1.15 light-years to get here. Therefore, it arrived at time 2.31 + 1.15 = 3.46.

Warning: You have not completely understood this example (and therefore have missed something important) unless you can explain where the numbers 1.73, 3.46, 2.31 and 1.15 came from!

16.3. Bob sees the video slowed down for two reasons: First, according to Bob, all earthbound clocks (and hence all earthbound streaming video servers) run slow. Second, according to Bob, earth is moving away from him, so the later parts of the movie have farther to travel than the earlier parts.

To calculate the exact rate of slowdown, first do everything in (earthbound) Alice's coordinates: Bob is traveling along the line $x = vt$. Alice, at some time t_0, sends him a movie frame that travels (at the speed of light) along the line $x = t - t_0$. To see when it arrives, solve $x = vt = t - t_0$ to get $t = t_0/(1 - v)$, $x = vt_0/(1 - v)$.

Now, Lorentz transform to Bob's coordinates. This gives

$$t' = \frac{t - xv}{\sqrt{1 - v^2}} = \frac{t_0(1 + v)}{\sqrt{1 - v^2}}$$

Therefore, the video arriving at Bob's ship runs slow by a factor of $(1 + v)/\sqrt{1 - v^2}$.

16.4.

This is just like the previous exercise, except that Bob's velocity is now $-v$, so the slowdown factor $(1 + v)/\sqrt{1 - v^2}$ becomes $(1 - v)/\sqrt{1 - v^2} = \sqrt{(1 - v)/(1 + v)}$. This is smaller than 1, so Bob actually sees the video speeded up.

16.5.

Now, Bob is no longer an inertial observer — he's constantly changing direction, so his frame is constantly changing, which means it would be very hard to think about this from Bob's point of view. Fortunately, it's easy to think about it from Alice's.

Alice says Bob is moving, hence Bob's clocks run slow. Therefore, if one of his orbits takes (say) an hour by Alice's watch, it will take only (say) 45 minutes by Bob's.

On this, *Bob must agree with Alice*, because they can both agree to note the times on their watches when Bob is directly over Alice's head, and then the next time he's directly over her head.

So, if Alice says Bob's clock runs slow, Bob must say Alice's runs fast. So, do all other earthbound clocks and video servers. Therefore, Bob sees the video speeded up.

There's an apparent paradox here: Alice is in motion with respect to Bob, so her clocks must run slow in Bob's frame. And they do. But Bob's frame keeps changing, and as we've seen before, a change in Bob's frame can appear to cause a jump forward in Alice's clock. After accounting for this effect, Alice's clock might appear to run either fast or slow. The argument above establishes that it in fact appears to run fast.

16.6.

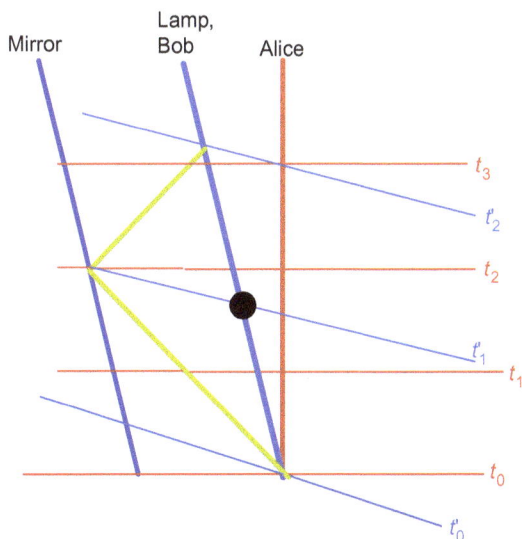

Alice's worldline is shown in boldfaced red, Bob's, the lamp's and the mirror's in boldfaced blue. The worldline of the light is shown in gold. The equally spaced horizontal red lines are Alice's lines of simultaneity and the equally spaced near-horizontal blue lines are Bob's lines of simultaneity.

Note that according to Alice, the outbound journey lasts from time t_0 to time t_2, but the return journey takes less time, lasting from t_2 to just a little past t_3.

Note that according to Bob, the outbound journey lasts from time t_0' to time t_1', and the inbound journey lasts from time t_1' to time t_2', which is equally long because the times are equally spaced.

To flesh this out a little more, suppose that Bob happens to scratch his nose at the event marked by the black circle. Then:

- Bob says: "At the moment the light beam hit the mirror, I was just scratching my nose. That happened at time t_1', exactly halfway between when the light left at t_0' and when it returned at t_2'.
- Alice says: "Halfway through the light beam's roundtrip journey, Bob scratched his nose. At that time, the light was still on its way to the mirror. A bit later, at a time more than halfway through the roundtrip journey, the light reached the mirror and turned around."

16.7. Here is the picture, showing everyone's worldlines, and (in gold) the worldline of Alice's light-signal:

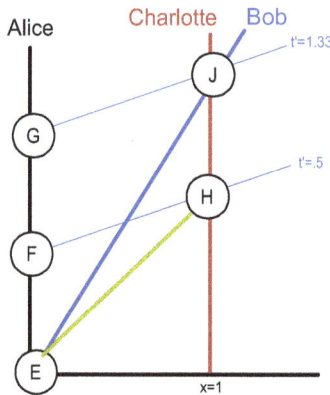

	x	t	x'	t'
E	0	0	0	0
F	0	2/5	-3/10	1/2
G	0	16/15	-4/5	4/3
H	1	1	1/2	1/2
J	1	5/3	0	4/3

Bob's (blue) lines of simultaneity have slope 3/5 in Alice's coordinates, which makes it easy to compute Alice's coordinates x and t for all the labeled pints. Then we use the Lorenz transformation to find Bob's coordinates in the last two columns of the chart.

Alice's story:

> **Time 0 (event E):** Bob passes me and turns on his stopwatch. I send my signal to Charlotte.
>
> **Time 1 (event F):** Charlotte receives my signal and turns on her stopwatch.
>
> **Time 1.67 (event J):** Bob and Charlotte meet and they turn off their stopwatches.
>
> Bob's watch ran for 1.67 seconds (from E to H) and advanced 1.33 seconds. It therefore ran at 4/5 speed.
>
> Charlotte's watch ran for .67 seconds (from G to H) and advanced .67 seconds. It therefore ran at normal speed.

Bob's story:

> **Time 0 (event E):** I cross paths with Alice. She sends out a signal and I start my stopwatch.
>
> **Time .5 (event G):** Charlotte receives the signal and turns on her stopwatch. Meanwhile, Alice's watch (at J) shows .4 because it's running at 4/5 speed.
>
> **Time 1.33 (event H):** I pass Charlotte and we turn off our watches.
>
> My watch ran for 1.33 seconds (from E to J) and advanced 1.33 seconds. It was running at normal speed.
>
> Charlotte's watch ran for .83 seconds (from H to J) and advanced .67 seconds. It was running at 4/5 speed.

16.8. Below we've depicted the worldlines of Alice and her clocks as vertical. We could equally correctly have depicted Bob's worldline as vertical, but it turns out that there's no need to draw Bob's worldline at all for this problem:

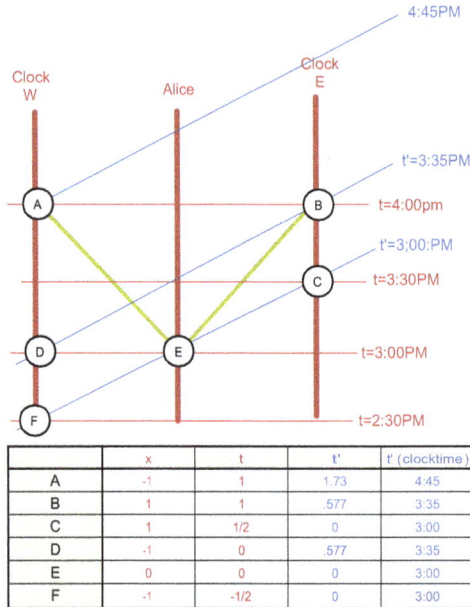

	x	t	t'	t' (clocktime)
A	-1	1	1.73	4:45
B	1	1	.577	3:35
C	1	1/2	0	3:00
D	-1	0	.577	3:35
E	0	0	0	3:00
F	-1	-1/2	0	3:00

Alice sends her signals at F, which we treat as the origin. Bob's (blue) lines of simultaneity all have slope $1/2$ in Alice's coordinates, which makes it easy to compute Alice's coordinates x and t for all the labeled points. Then we use the Lorentz transformation to find Bob's time coordinate t' for each point. In the last column above, we translate that time coordinate to a clock time, remembering that time 0 is 3:00 PM and there are sixty minutes in an hour. (We've rounded to the nearest five minutes.)

Alice's chronology:

3:00: I passed Bob and sent my signals.

4:00: The clocks received my signals and stopped.

Bob's chronology:

3:00: Alice passed me and sent her signals. The west clock was showing 2:30 and east clock was showing 3:30.

3:35: The east clock received its signal and stopped, showing 4:00. It had therefore advanced 30 minutes over the course of 35 minutes, so it was running at about 6/7 speed. (Bob says "about" because of rounding errors.)

4:45: The west clock received its signal and stopped, showing 4:00. It had therefore advanced 90 minutes over the course of 105 minutes., so it was running at about 6/7 speed.

Moral of the story: If Alice says that both clocks *showed the same time when they stopped*, Bob must agree. If Alice says that both clocks *stopped at the same time*, Bob is free to say otherwise.

17.1.

$$\sqrt{(\Delta t')^2 - (\Delta x')^2} = \sqrt{\left(\frac{\Delta t - v\Delta x}{\sqrt{1-v^2}}\right)^2 - \left(\frac{\Delta x - v\Delta t}{\sqrt{1-v^2}}\right)^2}$$

$$= \sqrt{\frac{((\Delta t)^2 - 2v\Delta t\Delta x + v^2(\Delta x)^2) - ((\Delta x)^2 - 2v\Delta t\Delta x + v^2(\Delta t)^2)}{1-v^2}}$$

$$= \sqrt{\frac{((\Delta t)^2 - (\Delta x)^2)(1-v^2)}{1-v^2}}$$

$$= \sqrt{(\Delta t)^2 - (\Delta x)^2}$$

18.1. According to Alice they occur in the order J, E, F, H. According to Bob, they occur in the order F, E, J, H (because the picture indicates that H is farther above Jeter's line of simultaneity than J is). According to Jeter, they occur in the order F, H, E, J.

21.1. Here, is the picture:

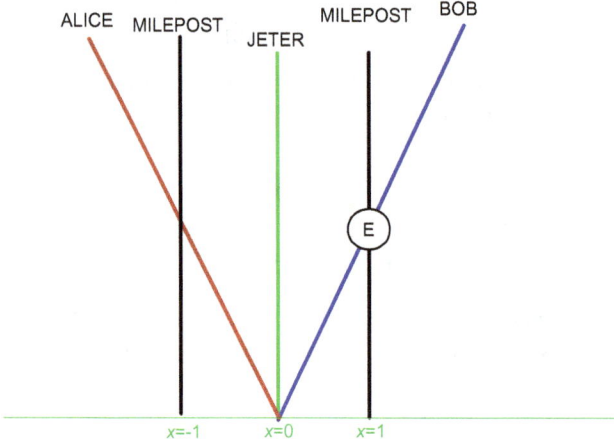

(a) Bob reaches the eastern milepost at event E. According to Jeter, Bob has traveled 1 light-year at speed $1/2$, so he must have arrived at time 2; that is, in Jeter's coordinates, E is at $x = 1, t = 2$.

(b) Bob is traveling at speed $v = 1/2$ with respect to Jeter, so we can use the Lorentz transformation to get Bob's coordinates for E: $x' = 0, t' = \sqrt{3}$. Thus, Bob says he reaches the milepost at time $\sqrt{3}$ (approximately 1.73).

(c) Alice is traveling at speed $v = -1/2$ with respect to Jeter, so we can use the Lorentz transformation, applied to $x = 1, t = 2$, to get Alice's coordinates for E: $x'' = 4/\sqrt{3}, t'' = 5/\sqrt{3}$. So, Alice says that Bob reaches the milepost at time $5/\sqrt{3}$ (approximately 2.89).

(d) From parts (b) and (c), Alice says that Bob reaches his milepost at time $5/\sqrt{3} \approx 2.89$, and that his clock reads $\sqrt{3} \approx 1.73$ when he gets there. So, his clock must be running at $\sqrt{3}/(5/\sqrt{3}) = 3/5$ of normal speed.

(e) Jeter is traveling westward at speed $v = 1/2$ with respect to Alice and Bob is traveling westward at speed $w = 1/2$ with respect to Jeter. So, to get Bob's speed with respect to Alice, use the velocity addition formula: $(v + w)/(1 + vw) = 1/(5/4) = 4/5$.

(f) From (e), Bob is traveling at speed $v = 4/5$ with respect to Alice. From Exercise 13.2, she must say his clock is running at a fraction $\sqrt{1 - v^2} = 3/5$ of normal speed. This matches what we found in part (d).

22.1.

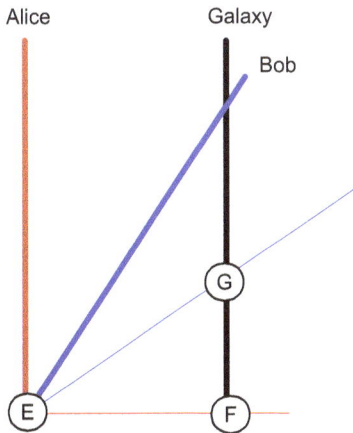

When Bob passes Alice at E, she defines the distance to "the galaxy now" as the spatial distance (in her frame) from E to F. He defines the distance to "the galaxy now" as the spatial distance (in his frame) from E to G.

We're told that Alice says that the distance to the galaxy now is D, so she assigns the coordinates $(x = D, t = 0)$ to event F.

Bob's line of simultaneity through G has slope v, so its coordinates (according to Alice) are $(x = D, t = vD)$.

Bob therefore assigns G the spatial coordinate

$$x' = (D - v^2 D)/\sqrt{1 - v^2} = D\sqrt{1 - v^2}$$

This, then, is what he says is the distance to the galaxy now.

22.2. The picture below shows Alice's worldline in red, Andromeda's in black, Bob's in blue, and the front of his 2.5-milion light-year long ruler in green. The dashed green line is the worldline of the 20 light-year mark on Bob's ruler (so the picture is very definitely not to scale).

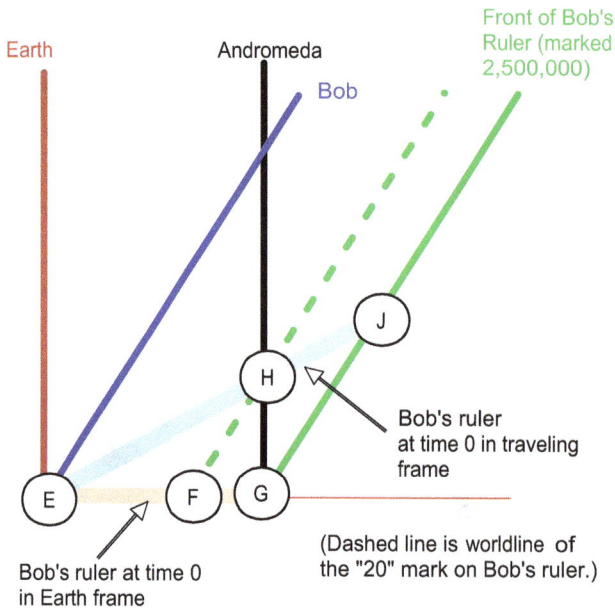

In the earth frame, Bob's ruler at time 0 is given by the pink line stretching from *E* to *G*. In Bob's frame (once he is traveling) his ruler at time 0 is given by the blue line stretching from *E* to *J*.

Just before Bob takes off, he says that the tip of his ruler just touches Andromeda at *G*. Just after he takes off, he says that his ruler extends well past Andromeda (out to *J*) and that the 20 light-year mark on his ruler is just touching Andromeda at *H*.

In neither frame does Bob claim that Andromeda has suddenly jumped. He does say something like "A moment ago I was using one frame, and Andromeda was 25 million light-years away; now I'm using a different frame where Andromeda is only

20 light-years away". But this is of course because Bob has changed his direction in spacetime; not because Andromeda has moved.

The laws of physics allow Bob to travel from here to the point "20" on his ruler in just over 20 years. We can hope he enjoys his journey.

At first, the 2.5 million mark on Bob's ruler just touches Andromeda. Immediately after he takes off, Alice says that the 20 mark on his ruler will *eventually* reach Andromeda, while Bob says it's already there. But of course Bob does not actually *observe* that it's already there; the light from Andromeda needs time to get to him. So, he *observes* no change at all, but when he arrives at Andromeda in 20 years, he can piece his story together after the fact.

23.1. Here is the picture:

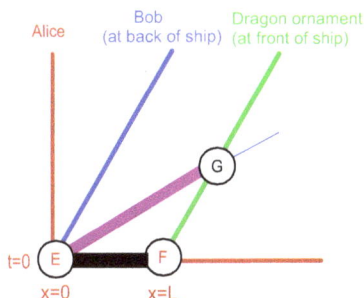

Bob's line of simultaneity through E and G has the equation $t = vx$. The worldline of the dragon passes through $(x = L, t = 0)$ and has slope $1/v$, so its equation is $x = vt + L$. Solving simultaneously gives the coordinates of G in Alice's frame:

$$x = \frac{L}{1 - v^2} \qquad t = \frac{Lv}{1 - v^2}$$

Inserting these into the Lorentz transformation gives

$$x' = \frac{L}{\sqrt{1 - v^2}} \qquad t' = 0$$

So as he passes Alice at the origin, Bob says that the front of his ship is at $x' = L/\sqrt{1-v^2}$ and therefore his ship has length $L/\sqrt{1-v^2}$.

23.2. Now the picture is this:

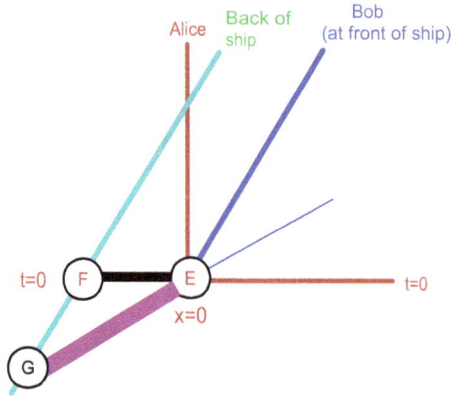

As Bob (at the front of his ship) passes Alice at E, she says that Bob's ship at this moment consists of the black line segment, while Bob says that his ship at this moment consists of the purple line segment.

Bob's line of simultaneity through the origin has the equation $t = vx$. The worldline of the back of the ship passes through the event F with coordinates $(x = -L, t = 0)$ and has slope $1/v$, so its equation is $tv = x + L$. Solving simultaneously gives the coordinates of G in Alice's frame:

$$x = -\frac{L}{1-v^2} \qquad t = -\frac{Lv}{1-v^2}$$

Inserting these into the Lorentz transformation gives

$$x' = -\frac{L}{\sqrt{1-v^2}} \qquad t' = 0$$

so that Bob says his ship has length $L/\sqrt{1-v^2}$, just as if he'd been sitting in the front of the ship.

23.3. Nice try, Alice, but it won't work. Here is the picture showing the (red) worldlines of the front and back of the tunnel and the (blue) worldlines of the front and back of the ship:

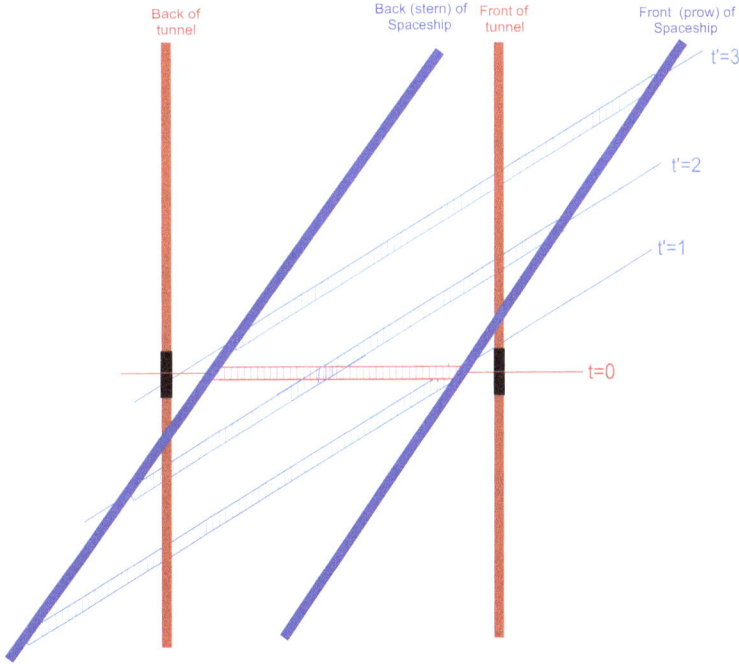

The black segments of the red worldlines represent the brief periods when the doors are closed.

The horizontal red striped line is Bob's ship at time 0 according to Alice.

The blue striped lines are Bob's ship at various times according to Bob.

Alice says: At time $t = 0$, both doors were closed and Bob's ship was fully inside the tunnel. That proves the tunnel is longer than the ship.

Bob says: At time $t' = 2$, both doors were open and the ends of my ship were sticking out of both ends of the tunnel. That proves the ship is longer than the tunnel.

Bob's story in more detail:

At $t' = 1$, my prow had entered the tunnel and was approaching the (closed) front door.

At $t' = 2$, the front door had opened in time for my prow to pass through. My stern was still sticking out the (open) back door.

At $t' = 3$, my stern had entered the tunnel and the back door had closed behind me.

25.1. This is a trick question. "They simultaneously drop rocks out the train's window" is ambiguous. Do Alice and Bob drop their rocks simultaneously according to Alice and Bob, or do they drop them simultaneously according to their ground-based friend Jeter? We'll answer both ways.

The diagram below shows the worldlines of Jeter (in black), Alice (in red) and Bob (in blue). We might as well have Alice drop her ball at point A, which we take to be the origin for both Alice and Jeter.

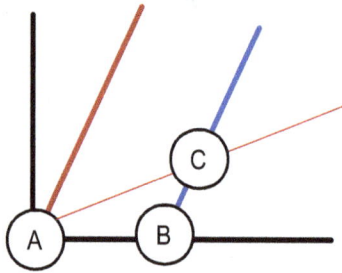

The thin red line is a line of simultaneity for Alice and Bob.

If Alice and Bob drop their rocks simultaneously according to Jeter (that is, at A and B) then the distance between them is (according to Jeter) equal to the length of the train, which is (again according to Jeter) $100\sqrt{1 - .7^2} \approx 71.4$ light-seconds.

If they drop their rocks simultaneously according to themselves (that is, at A and C), then we'll need the coordinates of those points.

In Alice's frame, the coordinates of those points are:

$$A: \quad x' = 0 \quad t' = 0$$
$$C: \quad x' = 100 \quad t' = 0$$

Transforming to Jeter's frame (and remembering that Jeter's velocity relative to Alice is $-.7$) we get

$$A: \quad x = 0 \quad t = 0$$
$$C: \quad x \approx 140 \quad t \approx 98$$

Therefore Jeter says the rocks are dropped about 140 light-seconds apart.

We can check our work by making sure the numbers make sense: According to Jeter, Bob is seated 71.4 light-seconds in front of Alice. Alice drops her rock. Then 98 seconds later, Bob drops his rock. During that 98 seconds, Bob hurtles forward at speed .7, so he travels $.7 \times 98 \approx 68.6$ light-seconds. Therefore his drop occurs about $71.4 + 68.6 \approx 140$ light-seconds in front of Alice's.

How You Could Have Discovered the Lorentz Transformations

Alice says Bob is running past her, at velocity v. (Bob, of course, begs to differ and insists that he is standing still.) They cross paths at an event that they both call the origin — that is, they label it as taking place at "time zero" and "location zero". A car crash occurs at a time and place that Alice calls t and x. When and where does Bob say the car crash takes place?

In Chapter 9, I gave you the answer. It is found by using the *Lorentz transformation*:

$$t' = \frac{t - vx}{\sqrt{1 - v^2}} \tag{9.1}$$

$$x' = \frac{x - vt}{\sqrt{1 - v^2}} \tag{9.2}$$

You might have wondered where those equations came from or how you might have discovered them on your own. This appendix will address those questions. We'll proceed in a series of steps.

Although we've frequently described Equations (9.1) and (9.2) as *transformations* (note the plural), it is sometimes more convenient to call the two equations, taken together, a single *transformation*. That's what we'll do in this appendix. This is strictly a vocabulary choice and has no further implications.

Step One: Applying the First Principle of Relativity. In Chapter 7, we met the First Principle of Relativity: Bob has exactly as much right as Alice to consider himself stationary. Nothing in

Bob's experience can force him to conclude that he is moving. We interpret this to mean that Bob and Alice must agree on the basic laws of physics.

For example: Cars generally speed up when you hit the accelerator and slow down when you hit the brake. That's how the world looks to Alice, so it should also be the way the world looks to Bob.

Conclusion:

> **Bob and Alice must agree on which cars (or any other objects) are speeding up, which are slowing down, and which are maintaining constant speeds.**

But maintaining a constant speed is the same thing as having a straight worldline. So, we can reword that conclusion:

> **Bob and Alice must agree on which worldlines are straight.**

Or, in just slightly different words:

> **The Lorentz transformation must take straight worldlines to straight worldlines.**

Since every timelike line is at least potentially the worldline of *something*, we can reword this once more:

> **The Lorentz transformation must take all timelike lines to timelike lines.**

Step Two: The General Form of the Lorentz Transformation. Here's, what we know:

I. By assumption, Alice and Bob have both labeled the same event as the origin.[1] Therefore, the Lorentz transformation must take the coordinates $(t = 0, x = 0)$ to the coordinates $(t' = 0, x' = 0)$.

---

[1]Without this upfront assumption, the Lorentz transformation would take the alternative form given by Equations (12.1) and (12.2).

(We express this by saying that the Lorentz transformation must *preserve the origin*.)

II. The Lorentz transformation must convert timelike lines to timelike lines. (This was the conclusion of Step One.)

One way to satisfy both requirements is to assume that the Lorentz transformation is *linear*, which means it has the form

$$t' = At + Bx$$
$$x' = Ct + Dx$$

$$(A.1)$$

where A, B, C and D are constants. Linear transformations are simple, easy to work with, and ubiquitous in mathematics. Better yet, it's quite easy to check that they always preserve the origin and always take lines to lines. If you choose the constants carefully, they even take *timelike* lines to *timelike* lines.

In fact, it turns out that if you want to satisfy requirements **I.** and **II.**, then linear transformations are the only candidates. It is a mathematical fact that no other transformation has these properties. Although that's a fact, it's not an obvious one, and seems to be remarkably little-known. The author of your book has directly verified that many topnotch mathematicians are unaware of it.

So, if you happen to know more than the average top-notch mathematician, then you realize that the Lorentz transformation *must* be linear. Otherwise, you can still reason this way:

"I need a transformation that preserves the origin, and takes timelike lines to timelike lines. The only good transformations I can think of are linear. So, I'll see if I can find a linear transformation that works. Of course there might be some other candidates that *aren't* linear, but since I can't figure out what those are, I'll just ignore them. I hope this doesn't lead me astray."

And fortunately you *won't* be led astray, because those other candidates do not in fact exist.

Conclusion:

> **The Lorentz transformation is linear —
> that is, it has the form of the Equations
> (A.1).**

Now all that remains is to figure out the values of the constants A, B, C and D.

Step Three: The Equation for Bob's Worldline. We've assumed that, according to Alice, Bob travels at velocity v and his worldline passes through the origin. This means that in her coordinates, the equation of his worldline is

$$x = vt$$

Of course in Bob's coordinates, where his velocity is 0 (and where he always occupies location 0), the equation of his worldline must be

$$x' = 0$$

Since these are two different descriptions of the same worldline, we can plug them simultaneously into the second equation of (A.1) to get

$$0 = Ct + Dvt$$

which gives

$$C = -Dv \qquad (A.2)$$

Step Four: Applying the Second Principle of Relativity. Consider a light beam passing through the origin with velocity 1. Its worldline in Alice's coordinates is therefore

$$x = t$$

The Second Principle of Relativity, which we encountered in Chapter 8, tells us that Bob and Alice must agree on the speed of light. Therefore, Bob agrees that the light beam has velocity 1, and its worldline in his coordinates is therefore

$$x' = t'$$

Since these are two different descriptions of the same worldline, we can plug them simultaneously into (A.1) to get

$$At + Bt = t' = x' = Ct + Dt$$

from which we get

$$A + B = C + D \qquad (A.3)$$

We can make the same argument for a light beam traveling the opposite direction, with a worldline given in Alice's coordinates by

$x = -t$ and in Bob's by $x' = -t'$. This gives us

$$A - B = -(C - D) \tag{A.4}$$

Step Five: Combining the Equations. Here, again are the three Equations (A.2), (A.3) and (A.4):

$$C = -Dv \quad A + B = C + D \quad A - B = -(C - D)$$

One solution to this system of equations is $A = B = C = D = 0$, which would clearly be absurd. (It would imply that in Bob's coordinates, every event occurs at the same time $t' = 0$ and the same location $x' = 0$). Discarding that solution, the only remaining solutions all have the form

$$D = A \quad B = C = -Av$$

Inserting this into (A.1), we have

$$t' = At - Avx$$
$$x' = -Avt + Ax \tag{A.5}$$

All that remains now is to figure out the value of A.

Step Six: Bob's point of view. If we interchange the roles of Bob and Alice (so that we are seeking the Lorentz transformation that gives Alice's coordinates (t, x) in terms of Bob's coordinates (t', x'), and if we remember that Alice's velocity with respect to Bob is not v but $-v$, then the same reasoning that led to Equations (A.5) leads us to

$$t = A't + A'vx$$
$$x = A'vt + A'x \tag{A.6}$$

for some possibly different constant A'.

When Alice says that a car crash takes place at time t and location x, Bob says it takes place at time t' and time x'. Equations (A.5) give t' and x' in terms of x and t. Equations (A.6) give t and x in terms of t' and x'. Combining all these equations, you can easily discover that

$$AA'(1-v) = 1 \tag{A.7}$$

Unfortunately, Equation (A.7) doesn't allow us to solve for A, because we don't know the value of A'. But we're almost there! First, we need a brief digression on time dilation.

Step Seven: Time Dilation: Alice's clock ticks (according to Alice) once per minute, first at time $t = 0$ and then at time $t = 1$. Both ticks occur at the location where Alice says she is standing still, which is $x = 0$.

Applying the first equation from the transformation (A.5) to ($t = 0, x = 0$), we see that according to Bob, the first tick takes place at time $t' = 0$. Applying the same equation to ($t = 1, x = 0$), we see that according to Bob, the second tick takes place at time $t' = A$. So, according to Bob, Alice's clock ticks every A minutes.

Likewise, according to Alice, Bob's clock ticks every A' minutes.

Step Eight: Mirroring the Universe. Did I forget to mention that Alice and Bob are live-blogging their adventures over Zoom? You and your cousin Jeter, who are both located some distance away and are stationary with respect to Alice, are watching on your monitors. Jeter, unlike you, has clicked the Zoom setting that says "Mirror my video." So the directions you call "left" and "right" are the directions Jeter calls "right" and "left". The velocity you call v is the velocity Jeter calls $-v$.

Therefore, while you employ the transformation (A.5) to get from Alice's coordinates to Bob's and (A.6) to get from Bob's coordinates to Alice's, Jeter does exactly the opposite.

Bob, visibly to both of you, has just recorded a diary entry saying "I see that Alice's clock is running slow." Since you've read Step Seven in this appendix, you will of course expect him to record that instead of ticking once per minute, Alice's clock is ticking A times per minute (where A is the mysterious constant in Equations (A.5). Jeter, who is reading the same appendix but has switched v with $-v$, will expect Bob to say that instead of ticking once per minute, Alice's clock is ticking A' times per minute. But of course you both see the same diary entry (though Jeter has to struggle a little to read it, because he sees it as a mirror image). Therefore, if the universe works the way you both expect it to, it must be the case that

$$A = A' \tag{A.8}$$

The reasoning in Step Eight is not ironclad. We could conceivably live in a universe that behaves very differently when seen in a mirror — maybe a universe where gravity pulls only to the left and not to the right, or where electrons repel protons on their left but attract protons on their right — or a universe where the Lorentz transformation works in a fundamentally different way depending on whether Bob is traveling leftward or rightward. But everyday experience suggests that the universe we live in is not quite so odd as that.

Step Nine: Putting it All Together. Combining Equations (A.7) and (A.8) you can deduce that

$$A = A' = \frac{1}{\sqrt{1 - v^2}}$$

and plugging this into the transformation (A.5), you'll finally discover that Equations (9.1) and (9.2) are the right equations!

www.ingramcontent.com/pod-product-compliance
Lightning Source LLC
Chambersburg PA
CBHW050629190326
41458CB00008B/2193